DEDUCTIVE GEOMETRY

DECIPHERING THE PROOF
A COMPREHENSIVE SOLUTION GUIDE

VOLUME 1

Raymond Guyamier

Order this book online at www.trafford.com
or email orders@trafford.com

Most Trafford titles are also available at major online boo k retailers.

Printed in the United States of America.

ISBN: 978-1-4907-0671-9 (sc)
ISBN: 978-1-4907-0672-6 (e)

Trafford rev. 06/24/2013

 www.trafford.com

North America & international
toll-free: 1 888 232 4444 (USA & Canada)
phone: 250 383 6864 ♦ fax: 812 355 4082

TABLE OF CONTENTS

INTRODUCTION

Philosophy and Application of the Book

This book will be useful to improve solving techniques in Deductive Geometry. It requires a solid foundation and understanding of basic geometry already learned from a Geometry textbook. The exercise problems, as well as the lesson themselves, should have previously been studied. Justification and demonstration problems are the focus of this text.

Approach

Each chapter begins with a brief **Summary of Essential Theorems and Definitions**. Then, there is one **Example** explaining how to use the theorems and definitions to solve a problem. The third section is **Exercises**, which will provide the reader an opportunity to practice applying the theorems and definitions. To gain the most from the Exercises, they should be completed before reading the solutions. The last section is the detailed **Solutions** of the Exercises.

The common steps used to find the solution are:
1. Draw an exact figure of what is to be proven.
2. Accurately find out, and list, the given parts of the problem.
3. Then list the Conclusion to be proven.
4. Mark the figure according to what you can deduce about it from the information given.
5. Create the Demonstration section:
    a. Separate the page vertically into two columns.
    b. On the left side, write the answer in form of logical **steps**.
    c. On the right side, give the corresponding **reasons**.

Every step must have a reason. The reason can be a given fact, or confirmed by a property, definition or theorem. Furthermore, a reason used in the proof can also come from earlier steps in the proof itself.

You will find the solutions of all the problems in the section labeled "Solutions." Only theorems or definitions already learned in class can be used.

It is important to try to solve the problem first before reading the solution.
This will give you self-confidence, and you will progress at your own pace.

Deciphering the Proof is for students, parents, and teachers who need practice solving proofs in Geometry. Specifically, where Geometry is part of the 5^e curriculum in a French program, or for American students taking Geometry between Grades 8 and 10.

The book shows, step-by-step, how to reason and solve Geometry problems, by writing solutions in a clear, logical, and deductive sequence. This strategy is called, "modeling." Students learn, by imitating the method, and eliminating all the non-value adding verbiage that are distracting to the grader. By showing the core steps required to solve a problem, students avoid extraneous text and steps which make the solution difficult to follow, and difficult for the grader to evaluate with precision.

Teachers can use the material, in class, by showing partial solutions (of the reasoning or the proof), and asking the students to complete the other part.

The book should be used as a complement to a Geometry textbook. It is especially beneficial for average students with difficulties writing the solution to a problem in a logical deductive process. I would recommend the user of my book to, first, try to solve the problems entirely, before comparing with the step-by-step solutions following each chapter.

In over 50 years of teaching, in Canada and in Florida, I have not found anything that meets this need, and started writing this material for my own students over ten years ago. All the other books just provided a brief description, or some clues, as how to solve the problems. In this book, the detailed solution is provided, and the reasoning associated with each step, the way a teacher would like to read when grading a problem, for accurate evaluation. This is based on my experience teaching for the last 50 years, in Montreal, Canada (Régionale de Chambly, Régionale Ste. Croix), and in Miami, Florida (in the French Program of the International Studies Program at Carver Middle School in the Dade County Public School System). I hope you find this book to be an indispensable asset to teaching Geometry, and welcome any feedback.

Acknowledgements

Special thanks to my wife and family for their continuous support to make this project a reality. Liliana Agudelo and Liliana Rios-Cristancho, my co-workers and friends, who reviewed the final versions. Annabel Lehmann for drafting detailed drawings of the figures for every Example and Problem. Tracy Sachs for a creating an engaging cover which conveys my aspirations of simplifying deductive geometry for all. Last but, but not least, my students for the last 50 years, who have been subjected to my test and learn approach to refine this method, and the content of this book, to a point where I can share it with others.

CHAPTER 1
TRIANGLES

Summary of Essential Theorems and Definitions

Parallel and Perpendicular Lines:

1) If two lines are perpendicular to a third line, then they are parallel to each other.
2) If two lines are parallel to a third line, then they are parallel to each other.
3) If two segments are parallel, then their supports (the supporting line on which the segment lies) are parallel.

Perpendicular Bisector:

4) The perpendicular bisector is the line perpendicular at the midpoint of a segment.
5) The perpendicular bisector is formed by the set of points equidistant from the ending points of a segment.
6) Any point equidistant from the ending points of a segment is on its perpendicular bisector.
7) In an isosceles triangle, the bisector of the vertex angle is at the same time the altitude, the perpendicular bisector and the median of the triangle.
8) If in a triangle, the bisector of an angle is at the same time the altitude, or the perpendicular bisector, or the median, then it is a isosceles triangle.
9) The three perpendicular bisectors of a triangle intersect at a point called **the circumcenter.** (You only need the intersection of two perpendicular bisectors to determine this center)
10) Through a point, you can construct only one line perpendicular to a given line.
11) The altitude of a triangle is the perpendicular line of a side passing through the opposite vertex.

Altitude:

1) The altitude is the line perpendicular to a side, and which passes through the opposite vertex.
2) The three altitudes of a triangle intersect at the same point, called **the orthocenter.**

Angle Bisector:

1) The angle bisector is the ray which divides an angle into two congruent angles.
2) Any point on the angle bisector is equidistant from the two sides of the angle.
3) If a point is equidistant from the two sides of an angle, then it is on the angle bisector.
4) The three angle bisectors of a triangle intersect at the same point, called **the incenter**.

Median:

1) The median is the ray joining a vertex to the midpoint of the opposite side.
2) The three medians of a triangle intersect at the same point, called **the centroid**.

Example

If ABC is a triangle, then the perpendicular bisector of segments $\overline{AB}$ and $\overline{BC}$ intersect at D. What can you say about this triangle, if triangles ABD, DBC, and DAC have the same perimeter?

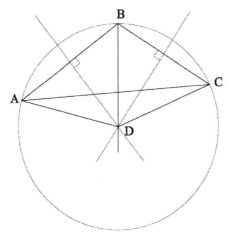

Given:

ABC is a triangle.

The perpendicular bisectors of segments $\overline{AB}$ and $\overline{BC}$ intersect at D.

Conclusion:

Find the exact nature of triangle ABC.

Demonstration:

Step	Reason
1) The perpendicular bisectors of segments $\overline{AB}$ and $\overline{BC}$ intersect at D.	Given
2) Then D is the circumcenter of triangle ABC.	The perpendicular bisectors of two sides of a triangle intersect at the circumcenter.
3) Let x be the measurement of segment $\overline{AD}$.	Assumption
4) AD = BD = CD = x	All are the radius of the circumscribed circle.
5) Perimeter of ABD = AD+BD+AB Or = x + x +AB Or = 2x + AB	Substituting AD and BD by their values.
6) Likewise Perimeter of BDC = BD+DC+BC	Substituting BD and DC by their values.

Step	Reason
Or $\quad$ = x + x + BC Or $\quad$ = 2x + BC	
7) Likewise Perimeter of ADC = AD+DC+AC Or $\quad$ = x + x + AC Or $\quad$ = 2x + AC	Substituting AD and DC by their values.
8) Therefore we have 2x + AB = 2x + BC = 2x + AC 9) AB = BC = AC 10) **Finally, triangle ABC must be equilateral.**	Given (the three triangles have the same perimeter). Subtraction property of equality (subtracting 2x from the equation). An equilateral triangle has three congruent sides.

Exercises

1.1 There are three triangles ABC, ABD, and ABE; with M, N, P as their respective circumcenters. Prove that M, N, and P are collinear.

1.2 Draw a circle with center O, and one of its chords $\overline{AB}$. Does the center (O) of the circle (*C*) lie on the perpendicular bisector of segment $\overline{AB}$?

1.3 Draw a triangle LMN such that LM = 6cm, LN = 5cm, and MN = 7cm. Then draw its circumscribed circle. The circumcenter is "C."
   a. Draw a triangle PQR such that P and Q are on the line $\overleftrightarrow{LN}$ and C is the circumcenter.
   b. Prove that the segments $\overline{PQ}$ and $\overline{LM}$ have the same midpoint.

1.4 Draw a triangle ABC such that AB = AC, and P is the midpoint of $\overline{BC}$. Let M be the circumcenter of triangle ABP, and N the circumcenter of triangle ACP. Prove that $\overleftrightarrow{MN}$ is parallel to $\overleftrightarrow{BC}$.

1.5 A triangle MUL has a perimeter of 13 cm, UM = 3cm, and LU = 5cm.
   a. Calculate ML, and draw the triangle MUL.
   b. Place the point E such that L is the midpoint of $\overline{UE}$.
   c. Where is the circumcenter of triangle MUE? Prove it.
   d. Prove that MUE is a right triangle.

Solutions

1.1 There are three triangles ABC, ABD, and ABE; with M, N, P as their respective circumcenters. Prove that M, N, and P are collinear.

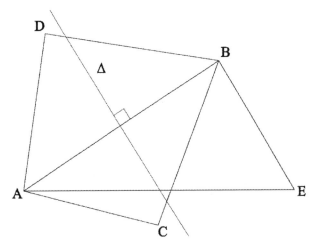

Given:

ABC, ABD, and ABE are three triangles.

M, N, and P are the respective circumcenters.

Conclusion:

Prove that M, N, and P are collinear.

Demonstration:

Step	Reason
1) In triangle ABC, M is on the perpendicular bisector of $\overline{AB}$.	The circumcenter is at the intersection of the three perpendicular bisectors of the triangle; hence on the perpendicular bisector (Δ) of $\overline{AB}$.
2) In triangle ABD, N is on the perpendicular bisector of $\overline{AB}$.	Same Reason as Step 1, $\overline{AB}$ is common to all three triangles.
3) In triangle ABE, P is also on the perpendicular bisector of $\overline{AB}$.	Same Reason as Step 1, $\overline{AB}$ is common to all three triangles.
4) **Therefore, M, N, and P are collinear.**	They are all three on the perpendicular bisector Δ of $\overline{AB}$.

1.2 Draw a circle C with center O and one of its chords $\overline{AB}$. Does the center (O) of the circle (**C**), lie on the perpendicular bisector of segment $\overline{AB}$?

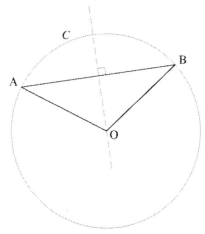

Given:

C is a circle with center O.

$\overline{AB}$ is a chord of C.

Conclusion:

Prove that the center, O, of the circle C, lies on the perpendicular bisector of segment $\overline{AB}$.

Demonstration:

Step	Reason
1) O is the center of the circle C. 2) Then OA = OB. 3) O is at the same distance from A and B.	Given Radii of the same circle are congruent. All the points on the circle are at same distance from the center.
4) **Therefore, O is on the perpendicular bisector of** $\overline{AB}$.	Any point equidistant from the ending points of a segment is on its perpendicular bisector.

1.3 Draw a triangle LMN such that LM = 6cm, LN = 5cm, and MN = 7cm. Then draw its circumscribed circle. The circumcenter is "C."

   a. Draw a triangle PQR such that P and Q are on the line $\overleftrightarrow{LM}$ and C is the circumcenter.

   b. Prove that the segments $\overline{PQ}$ and $\overline{LM}$ have the same midpoint.

<u>Study drawing</u>:

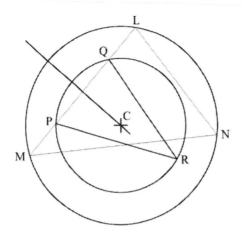

<u>Finding how to draw the figure</u>:

The problem is to place the points P and Q.

The circumcenter of PRQ must be C, which is the same as LMN.

C is at the intersection of the three perpendicular bisectors of LMN.

<u>Explanation of the figure</u>:

Draw the triangle LMN.

Draw two perpendicular bisectors of LMN; C is their intersection.

From C, draw a circle cutting $\overline{LM}$ at P and Q.

Place a third point R on the circle.

<u>Exact construction</u>:

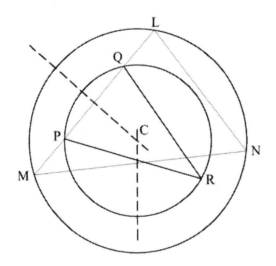

Given:

LMN is a triangle; C is its circumcenter.

PQR is a triangle.

P and Q are points on $\overline{LM}$.

C is also the circumcenter of triangle PQR.

Conclusion:

Prove $\overline{PQ}$ and $\overline{LM}$ have the same midpoint.

Demonstration:

Step	Reason
1) In the triangle LMN, C is on the perpendicular bisector of $\overline{LM}$.	The circumcenter of a triangle is the point of intersection of the three perpendicular bisectors.
2) Let K be the midpoint of $\overline{LM}$.	Assumption.
3) K is on the perpendicular bisector of $\overline{LM}$ passing through C.	The midpoint of a segment is on its perpendicular bisector.
4) Then $\overleftrightarrow{CK}$ is the perpendicular bisector of $\overline{LM}$.	Two points determine a line.
5) C is the circumcenter of triangle PQR.	Given.
6) Then C is on the perpendicular bisector of $\overline{PQ}$.	Same as in Step 1.
7) Let K' be the midpoint of $\overline{PQ}$.	Assumption.
8) K' is on the perpendicular bisector of $\overline{PQ}$ passing through C.	The midpoint of a segment is on its perpendicular bisector.
9) Then $\overleftrightarrow{CK'}$ is the perpendicular bisector of $\overline{PQ}$.	Two points determine a line.
10) Then K and K' are the same point.	Through a point, you can construct only one line perpendicular to a given line.
11) **Therefore, $\overline{LM}$ and $\overline{PQ}$ have the same midpoint.**	K and K' are the same point, and the midpoint of $\overline{LM}$ and $\overline{PQ}$.

1.4 Draw a triangle ABC such that AB = AC, and P is the midpoint of $\overline{BC}$. Let M be the circumcenter of triangle ABP, and N the circumcenter of triangle ACP. Prove that $\overleftrightarrow{MN}$ is parallel to $\overline{BC}$.

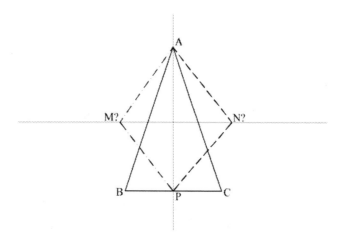

Given:

ABC is a triangle with AB = AC.

P is the midpoint of $\overline{BC}$.

M is the circumcenter of triangle APB.

N is the circumcenter of triangle APC.

Conclusion:

Prove $\overleftrightarrow{MN}$ is parallel to $\overleftrightarrow{BC}$.

Demonstration:

Step	Reason
1) AB = AC Then ABC is an isosceles triangle	Given. An isosceles triangle is a triangle with two congruent sides.
2) P is the midpoint of $\overline{BC}$	Given.
3) Then $\overline{AP}$ is the median of the base.	The median of a triangle is the line connecting the vertex to the midpoint of its opposite side.
4) And $\overline{AP}$ is the altitude for $\overline{BC}$.	In an isosceles triangle, the median of the base is also its altitude.
5) **Then $\overline{AP}$ is perpendicular to $\overline{BC}$.**	The altitude is the perpendicular line connecting a vertex to the opposite side.
In the triangle ABP 6) MA = MP	In a triangle, the circumcenter is at the same distance from the three vertices.
7) Then M is on the perpendicular bisector of $\overline{AP}$	Any point equidistant from the ending points of a segment is on its perpendicular bisector.
In the triangle APC 8) NA = NP	(Same Reasons as Steps 6 and 7) In a triangle, the circumcenter is at the same distance from the three vertices.
9) Then N is on the perpendicular bisector of $\overline{AP}$.	Any point equidistant from the ending points of a segment is on its perpendicular bisector.
10) $\overleftrightarrow{MN}$ is the perpendicular bisector of $\overline{AP}$.	M and N are both on the perpendicular bisector of $\overline{AP}$, and two points determine a line.
11) Therefore $\overline{AP}$ is perpendicular to $\overleftrightarrow{MN}$.	The perpendicular bisector of a segment is the perpendicular line on the midpoint of

Step	Reason
	this segment.
12) **Therefore $\overleftrightarrow{MN}$ is parallel to $\overleftrightarrow{BC}$.**	If two lines ($\overleftrightarrow{MN}$ and $\overleftrightarrow{BC}$) are perpendicular to a third one ($\overleftrightarrow{AP}$), then they are parallel.

1.5 A triangle MUL has a perimeter of 13 cm, UM = 3cm, and LU = 5cm.
   a.   Calculate ML, and draw the triangle MUL.
   b.   Place the point E such that L is the midpoint of $\overline{UE}$.
   c.   Where is the circumcenter of triangle MUE? Prove it.
   d.   Prove that MUE is a right triangle.

Calculations:

Perimeter MUL = 13cm	Given
Perimeter MUL = UM + LU + ML	Definition of perimeter.
Or 13 = 3 + 5 + ML	Substitution Property.
And ML = 13 – 8	Subtraction Property of Equality (subtracting 8 from the equation).
ML = 5cm	

Drawing:

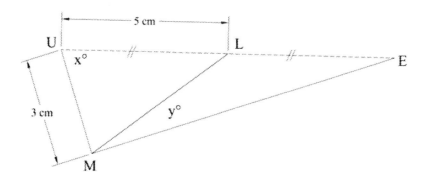

Given:
MUE is a triangle such that UM = 3cm; LU = 5cm; ML = 5cm.
L is the midpoint on $\overline{UE}$.
Conclusion:
Find the circumcenter of triangle MUE.

Demonstration:

Step	Reason
1) If LU = LE = 5cm	Construction of the midpoint of $\overline{UE}$.
2) Then the circle with center L passes through U and E.	$\overline{LU}$ and $\overline{LE}$ are radii (= 5cm).
3) ML = 5cm	See Calculation for Question a.
4) Then M is also on the circle with center L.	Same length from L (radius = 5cm).
5) **Therefore, L is the circumcenter of triangle MUE.**	The circle passing through the three vertices of the triangle.
Prove that UME is a right triangle. 1) Let $x°$ be the measurement of angle ∠MUL.	Assumption.
2) MUL is an isosceles triangle.	A triangle with two congruent sides (LU = LM = 5cm) is an isosceles triangle.
3) Then ∠ UML = $x°$.	Base angles of an isosceles triangle are congruent.
4) Let $y°$ be the measurement of angle ∠LME.	Assumption.
5) LEM is an isosceles triangle	A triangle with two congruent sides (LM = LE = 5cm) is an isosceles triangle.
6) Then ∠LEM = $y°$.	Base angles of an isosceles triangle are congruent.
7) Therefore, in the triangle UME we have: a. $x° + x° + y° + y° = 180°$	The sum of the measures of the angles of a triangle equals 180°.
b. Or $2x + 2y = 2(x + y) = 180°$	Reducing terms.
c. And $x + y = 90°$	Division Property of Equality (dividing by both sides of the equation by 2).
d. Then ∠UME = 90°	
8) **Finally, UME is a right triangle.**	A right triangle is a triangle with a right angle.

CHAPTER 2
CENTRAL SYMMETRY

Summary of Essential Theorems and Definitions

Central Symmetry (or Point Symmetry): Symmetry with respect to a point.

1) The symmetric point of point A, with respect to point M (the center of symmetry), is point A', such that M is the midpoint of $\overline{AA'}$. The abbreviation for A' being the symmetric point of point A, with respect to M, is written as follows: A' = $\mathbf{S}_M$ (A)

2) The symmetric segment of a segment, with respect to a point, is the segment formed by joining the symmetric points of the two endpoints of the segment.

3) Two symmetric segments have the same measurement.

4) The symmetric segment (of a segment) is a parallel segment.

5) The symmetric point of the midpoint of a segment is the midpoint of its symmetric segment.

6) The symmetric line (of a line) is a parallel line.

7) The symmetric angle (of an angle) is the angle of its symmetric sides (rays).

8) The symmetric angle (of an angle) is an angle with the same angle measurement.

9) The symmetric circle (of a circle) is a circle with the same radius or diameter, and its center is the symmetric point of the center of the original circle.

10) The endpoints of the diameter of a circle are symmetric points with respect to the center of the circle.

11) The symmetric triangle (of a triangle) is a triangle formed with the symmetric points of its vertices.

12) If a point is on the center of symmetry, then it is the symmetric point of itself.

Axial Symmetry: Symmetry with respect to a line or axis

Same properties as Central Symmetry **EXCEPT:**

1) The symmetric line of a line with respect to an axis **IS NOT A PARALLEL LINE**.

2) Two points A and B are symmetric, with respect to an axis Δ, if Δ is the perpendicular bisector of segment $\overline{AB}$.

3) If a point is on the axis of symmetry, then it is the symmetric point of itself.

$\mathbf{S(\ )}$ *is the symbol for "the symmetric of ..."*

Example

Draw an isosceles triangle ABD such that AB = BD = 4cm, and $\overline{AB}$ is perpendicular to $\overline{BD}$.
Draw the circle **C** with diameter $\overline{AB}$ and center O.
The points P, J, and L are the symmetric points of B, O, and A with respect to point D.
Draw the symmetric circle **C'** of circle **C** with respect to point D.

    a.   Determine its center and its radius
    b.   What is the measurement of $\overline{PL}$?
    c.   Is $\overleftrightarrow{PJ}$ perpendicular to the line $\overleftrightarrow{BD}$? Prove it.
    d.   What is the symmetric line of $\overleftrightarrow{BL}$ with respect to D?
    e.   What can be deduced for lines $\overleftrightarrow{BL}$ and $\overleftrightarrow{PA}$?

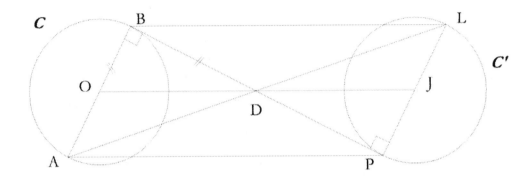

Given:

ABD is an isosceles triangle; AB = BD = 4cm.
$\overleftrightarrow{AB}$ is perpendicular to $\overleftrightarrow{BD}$.
C is a circle with diameter $\overline{AB}$, and O is its center.
P, J, and L are symmetric points of B, O, and A with respect to D.
C' is the symmetric circle of **C** with respect to D.

Conclusion:

Find the center and radius of **C'**.

Demonstration:

Step	Reason
1) If **C'** is the symmetric circle of C, then its center point must be symmetric with respect to D.	The symmetric circle (of a circle) is a circle with the same radius, and its center is the symmetric point of the original circle.
2) J is the symmetric point of O with respect to D.	Given.
3) **Then the center of C' is J, and the radius of C' is 2cm.**	AB ÷ 2 = 4cm ÷2 = 2cm
What is the measurement of $\overline{PL}$?	

Step	Reason
4) PL = 2 radii	1 diameter = 2 radii
5) **PL = 2 x 2cm = 4 cm**	
Is $\overleftrightarrow{PJ}$ perpendicular to the line $\overleftrightarrow{BD}$?	
6) B and P are symmetric with respect to D.	Given.
7) BD = DP	If two points (B and P) are symmetric with respect to a point (D), then that point (D) is the midpoint of the segment ($\overline{BP}$).
8) OD = DJ	Same Reason as Step 7.
9) BO = PJ	The radii of two symmetric circles have the same measurement.
10) Therefore, triangles BOD and DPJ are congruent.	Triangles with three congruent sides are congruent.
11) ∠OBD = ∠DPJ	Corresponding angles of congruent triangles are congruent.
	$\overleftrightarrow{AB}$ is perpendicular to $\overleftrightarrow{BD}$ (Given).
12) **If ∠OBD = 90°, then ∠DPJ = 90° and $\overleftrightarrow{PJ}$ is perpendicular to $\overleftrightarrow{BD}$.**	The sides of a right angle are perpendicular.
What is the symmetric line of $\overleftrightarrow{BL}$ with respect to D?	
13) P is the symmetric point of B with respect to D.	Given
14) A is the symmetric point of L with respect to D.	Given
15) **Thus $\overleftrightarrow{AP}$ is the symmetric line of $\overleftrightarrow{BL}$ with respect to D**	Two points determine a line.
What can be deducted for lines $\overleftrightarrow{BL}$ and $\overleftrightarrow{PA}$?	
16) **$\overleftrightarrow{BL}$ and $\overleftrightarrow{AP}$ are parallel**	If two lines are symmetric with respect to a point, then they are parallel.

Exercises

2.1 Draw an angle ∠xOy = 35° and a point B inside the angle. The point B is the symmetric point of A with respect to $\overrightarrow{Ox}$, and C is the symmetric point of B with respect to $\overrightarrow{Oy}$.

   a.   Prove that O is the center of a circle passing through B and C, and $\overline{OA}$ is its radius.

   b.   Prove that ∠AOC = 70°.

2.2 Draw a right triangle ABC, with $\angle$ABC = 90°, and I the midpoint of $\overline{AB}$; AB = 4cm, BC = 5cm, M an exterior point to ABC, such that AM = 4cm and MC = 3cm. Draw R, S, T, and J symmetric points of C, B, A and I with respect to M.

    a.   What is the symmetric segment of $\overline{BC}$ with respect to M?

    b.   What is the measurement of segment of $\overline{SR}$?

    c.   What is the symmetric segment of $\overline{RT}$ with respect to M?

    d.   Why are the lines $\overleftrightarrow{AC}$ and $\overrightarrow{RT}$ parallel?

    e.   What is the measurement of $\angle$RST?

    f.   What is the symmetric circle, with respect to M, of the circle with center I and diameter $\overline{AB}$?

2.3 Draw two intersecting circles **C** and **C'** with center A and B, and a common secant $\overleftrightarrow{KL}$. The segments $\overline{KN}$ and $\overline{LM}$ are diameters of **C;** the segments $\overline{KQ}$ and $\overline{LP}$ are diameters of **C'**. Demonstrate that lines $\overrightarrow{MN}$ and $\overrightarrow{PO}$ are parallel.

2.4 Draw a line d and two points, M and N, not on line d. The line $\overleftrightarrow{MN}$ is not parallel to d. Draw the line d', a symmetric line of d with respect to M. Draw the line d'', a symmetric line of d with respect to N. What can you say and prove about the position of d' and d''?

2.5 Draw a circle with diameter $\overrightarrow{AB}$ = 6cm, and center O. The perpendicular line to $\overrightarrow{AB}$ passing through O cuts the circle at O'. Let D be the midpoint of $\overline{OO'}$. Draw the symmetry of the entire drawing with respect to D. The two circles intersect at points I and J. Demonstrate that line $\overleftrightarrow{IJ}$ is the perpendicular bisector of segment $\overline{OO'}$.

2.6 Two segments $\overline{AB}$ and $\overline{CD}$ intersect at their midpoint M. AB = CD. $\overline{AB}$ and $\overline{CD}$ are **not** perpendicular. L is the symmetric point of A with respect **to the line** $\overrightarrow{MC}$. The lines $\overleftrightarrow{LC}$ and $\overrightarrow{BD}$ intersect at J.

    a.   Prove that the angles $\angle$MCA and $\angle$MDB are congruent.

    b.   Prove that triangle JDC is an isosceles triangle.

2.7. Draw a triangle ABC, and a point I, exterior to the triangle. Draw the symmetric points A', B', C' of A, B, C with respect to I (respectively). Demonstrate that the perpendicular bisectors of $\overline{AB}$ and $\overline{A'B'}$ are parallel.

Solutions

2.1 Draw an angle $\angle xOy = 35°$ and a point B inside the angle. The point B is the symmetric point of A with respect to $\overrightarrow{Ox}$, and C is the symmetric point of B with respect to $\overrightarrow{Oy}$.

   a. Prove that O is the center of a circle passing through B and C, and $\overline{OA}$ is its radius.

   b. Prove that $\angle AOC = 70°$.

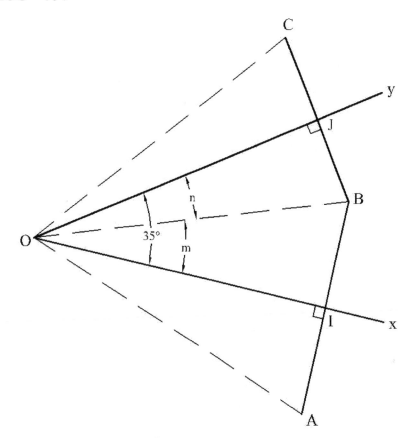

Given:

$\angle xOy = 35°$

$B = S_{\overline{Ox}}(A)$

$C = S_{\overline{Oy}}(B)$

Conclusion:

Prove that O is the center of a circle passing through B and C, and $\overline{OA}$ is the radius of the circle.

Demonstration:

Step	Reason
1) O is the symmetric point of itself with respect to $\overrightarrow{Ox}$.	If a point is on the axis of symmetry, then it is the symmetric point of itself.
2) $B = S_{\overline{Ox}}(A)$	Given
3) Then $\overline{OA}$ and $\overline{OB}$ are symmetric with respect to $\overrightarrow{Ox}$.	The symmetry of a segment is formed by joining the symmetry of two of its end points.
4) And OA = OB.	Two symmetric segments are congruent.

Step	Reason
5) OB = OC	Same reasons as Steps 1-4.
6) OA = OB = OC	Transitive property.
7) Then A, B, and C are the same distance from O.	
8) **O is the center of the circle with a radius of OA, and passing through B, and C.**	A circle is formed by all the points located at the same distance from a fix point, called the center of the circle
Prove that $\angle AOC = 70°$.	
9) I is the intersection of $\overrightarrow{AB}$ and $\overrightarrow{Ox}$.	Assumption.
10) I is the symmetric point of itself with respect to $\overrightarrow{Ox}$.	If a point is on the axis of symmetry, then it is the symmetric point of itself.
11) $\overline{OI}$ is the symmetric line of itself.	A line is formed by joining two points.
12) $\overline{OA}$ and $\overline{OB}$ are symmetric segments with respect to $\overrightarrow{Ox}$.	See Step 3.
13) Then $\angle AOI$ and $\angle IOB$ are symmetric.	The angles are formed by their respective symmetric sides.
14) And $\angle AOI = \angle IOB = m°$ (If m° is their common measurement).	If two angles are symmetric then they are congruent.
15) J is the intersection of $\overline{BC}$ and $\overrightarrow{Oy}$.	Given
16) $\angle BOJ = \angle JOC = n°$	Same Reasons as Steps 9-14, and n° is their common measurement.
17) $\angle IOJ = \angle xOy = 35°$	Given
18) $\angle IOJ = \angle IOB + \angle BOJ$	Adding angles.
19) $\angle IOJ = m° + n° = 35°$	Substituting angles by their values.
20) $\angle AOC = \angle AOI + \angle IOB + \angle BOJ + \angle JOC$	Adding the angles.
21) $\angle AOC = m° + m° + n° + n°$	Substituting the angles by their values.
22) $\angle AOC = 2(m° + n°) = 2(35°)$	$m° + n° = 35°$ (see Step 19).
23) **Therefore, $\angle AOC = 70°$.**	Basic multiplication.

2.2 Draw a right triangle ABC, with $\angle ABC = 90°$, and I the midpoint of $\overline{AB}$; AB = 4cm, BC = 5cm, M an exterior point to ABC, such that AM = 4cm and MC = 3cm. Draw R, S, T, and J symmetric points of C, B, A and I with respect to M.

   a. What is the symmetric segment of $\overline{BC}$ with respect to M?
   b. What is the measurement of segment of $\overline{SR}$?
   c. What is the symmetric segment of $\overline{RT}$ with respect to M?

d. Why are the lines $\overleftrightarrow{AC}$ and $\overleftrightarrow{RT}$ parallel?

e. What is the measurement of ∠RST?

f. What is the symmetric circle, with respect to M, of the circle with center I and diameter $\overline{AB}$?

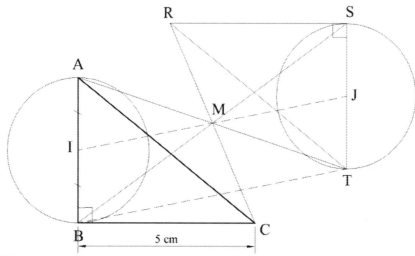

Given:

ABC is a right triangle, where angle ABC is the right angle

I is the midpoint of $\overline{AB}$

AB = 4cm; BC = 5cm

M is an exterior point to ABC

R, S, T, and J are symmetric points of C, B, A, and I with respect to M

Conclusion:

Find the symmetric segment of $\overline{BC}$ with respect to M.

Demonstration:

Step	Reason
1) S is the symmetric point of B, with respect to M.	Given
2) R is the symmetric point of C, with respect to M.	Given
3) **Then $\overline{SR}$ is the symmetric segment of $\overline{BC}$ with respect to M.**	The symmetric segment of a segment, with respect to a point, is the segment formed by joining its symmetric endpoints
What is the measurement of segment $\overline{SR}$?	
4) BC = 5cm	Given
5) **Then BC = SR = 5cm**	If two segments are symmetric, then they are

Step	Reason
	congruent.
What is the symmetric segment of $\overline{RT}$ with respect to M? 6) T and A are symmetric with respect to M. 7) C and R are symmetric with respect to M. 8) $\overline{AC}$ and $\overline{RT}$ are symmetric segments with respect to M.	Given Given Two segments are symmetric if their endpoints are symmetric.
Why are the lines $\overleftrightarrow{AC}$ and $\overleftrightarrow{RT}$ parallel? 9) $\overline{AC}$ is parallel to $\overline{RT}$ 10) **Therefore $\overleftrightarrow{AC}$ is parallel to $\overleftrightarrow{RT}$**	If two segments are symmetric with respect to a point, then they are parallel. If two segments are parallel, then their supports are parallel. See Chapter 1 Summary of Essential Theorems and Definitions #3.
What is the measurement of ∠RST? 11) A, B, and C are the symmetric points of T, S, and R (respectively). 12) Triangles ABC and RST are symmetric with respect to M. 13) The corresponding angles for triangles ABC and RST are congruent (∠ABC and ∠RST). 14) ∠ABC = 90° 15) **Therefore ∠RST = 90°**	Given The symmetric triangle (of a triangle) is a triangle formed with the symmetric points of its vertices. If two triangles are symmetric, then their corresponding angles are congruent. Given Transitive property.
What is the symmetric circle, with respect to M, of the circle with center I and radius IA? 16) The symmetric point of I with respect to M is J. 17) J is the midpoint of $\overline{ST}$. 18) $\overline{AB}$ and $\overline{ST}$ are symmetric with respect to M.	Given The symmetric midpoint of a segment is the midpoint of its symmetric segment. Corresponding sides of symmetric triangles are symmetric.

Step	Reason
19) $\overline{AB}$ and $\overline{ST}$ are congruent.	Symmetric segments are congruent.
20) $\overline{AB}$ is the diameter of the circle with center I. And $\overline{ST}$ is the diameter of the circle with center J.	The midpoint of the diameter of a circle is the center of the circle.
21) **Therefore, the symmetric circle of the circle with center I and diameter $\overline{AB}$ is the circle with center J and diameter $\overline{ST}$.**	The symmetric circle (of a circle) is a circle with the same radius or diameter, and its center is the symmetric point of the center of the original circle.

2.3 Draw two intersecting circles C and C' with center A and B, and a common secant $\overrightarrow{KL}$. The segments $\overline{KN}$ and $\overline{LM}$ are diameters of C; the segments $\overline{KQ}$ and $\overline{LP}$ are diameters of C'. Demonstrate that lines $\overleftrightarrow{MN}$ and $\overleftrightarrow{PQ}$ are parallel.

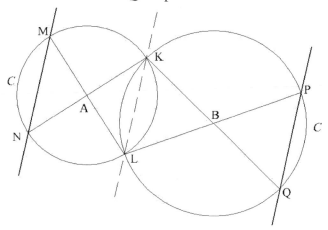

Given:

C and C' are circles with center A and B

$\overline{KN}$ and $\overline{LM}$ diameters of C

$\overline{KQ}$ and $\overline{LP}$ diameters of C'

Conclusion:

Prove that $\overleftrightarrow{MN}$ and $\overleftrightarrow{PQ}$ are parallel.

Demonstration:

Step	Reason
1) A, is the center of the circle **C**, and is the intersecting point of $\overline{KN}$ and $\overline{LM}$.	The diameters of a circle intersect at its center.
2) A is the midpoint of $\overline{KN}$.	The center of a circle is the midpoint of the diameter.

Step	Reason
3) N is the symmetric point of K with respect to A.	The endpoints of the diameter of a circle are symmetric points with respect to the center of the circle.
4) M is the symmetric point of L with respect to A.	Same reasoning as steps 1, 2, and 3 before with diameter $\overline{LM}$.
5) Lines $\overleftrightarrow{MN}$ and $\overleftrightarrow{KL}$ are symmetric lines with respect to A.	The symmetric line of a line, with respect to a point, is the line formed by joining the symmetric points of two of its points.
6) Therefore $\overleftrightarrow{MN}$ is parallel to $\overleftrightarrow{KL}$.	If two lines are symmetric with respect to a point, then they are parallel.
Prove $\overleftrightarrow{PQ}$ and $\overleftrightarrow{KL}$ are parallel lines 7) $\overleftrightarrow{PQ}$ and $\overleftrightarrow{KL}$ are symmetric lines with respect to B, and they are parallel.	Same reasoning as Steps 1-6, with diameter $\overline{LM}$, applied to diameter $\overline{LP}$.
8) **Therefore $\overleftrightarrow{MN}$ and $\overleftrightarrow{PQ}$ are both parallel to $\overleftrightarrow{KL}$, then they are parallel.**	If two lines are parallel to a third line, then they are parallel to each other. (See Chapter 1 Summary of Essential Theorems and Definitions #2).

2.4 Draw a line d and place two points, M and N, not on line d. The line $\overleftrightarrow{MN}$ is not parallel to d. Draw the line d', a symmetric line of d with respect to M. Draw the line d'', a symmetric line of d with respect to N. What can you say and prove about the position of d' and d''?

<u>Explanation of the figure:</u> lines d' and d''

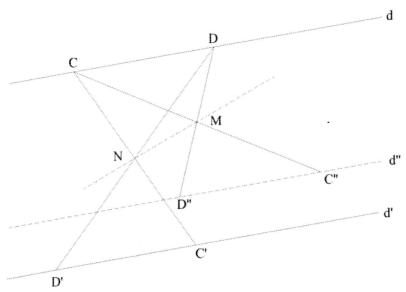

Choose any two points, C and D, on d (two points determine a line).
Then draw their symmetric points C' and D' with respect to M.
Draw line $\overleftrightarrow{C'D'}$ which is also the line d.
In the same way draw d', with N as the center of symmetry.

Given:

d is a line, M and N are points such that $\overrightarrow{MN}$ and d are not parallel

$d = S_M(d)$ d' $= S_N(d)$

Conclusion:

Find and prove the exact nature of lines d and d'.

Demonstration:

Step	Reason
1) $d = S_M(d)$	Given
2) d' is parallel to d.	If two lines are symmetric with respect to a point, then they are parallel.
3) d' $= S_N(d)$	Given
4) d'' is parallel to d.	If two lines are symmetric with respect to a point, then they are parallel.
5) **Therefore, d' is parallel to d'**	If two lines are parallel to the same third line, then they are parallel to each other.

2.5 Draw a circle with diameter $\overleftrightarrow{AB}$ = 6cm, and center O. The perpendicular line to $\overleftrightarrow{AB}$ passing through O cuts the circle at O'. Let D be the midpoint of $\overline{OO'}$. Draw the symmetry of the entire drawing with respect to D. The two circles intersect at points I and J. Demonstrate that line $\overleftrightarrow{IJ}$ is the perpendicular bisector of segment $\overline{OO'}$.

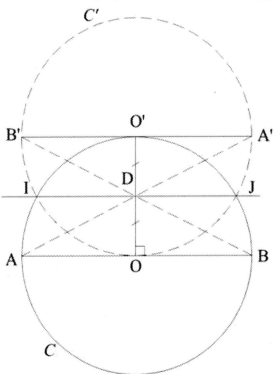

<u>Study Drawing:</u>

Draw A', B', and O' the symmetric points of A, B, O with respect to D.

Draw circle C', the symmetric circle of circle C with respect to D.

Given:

C is a circle with diameter AB = 6cm.

$\overline{OO'}$ is a perpendicular segment to $\overline{AB}$.

D is the midpoint of $\overline{OO'}$.

C and C' are symmetric circles with respect to D.

D is the center of symmetry.

Conclusion:

Prove that $\overleftrightarrow{IJ}$ is the perpendicular bisector of $\overline{OO'}$.

Demonstration:

Step	Reason
1) Let r be the length of the radius of circle C.	Assumption
2) The radius of C' is also r.	If two circles are symmetric, then they have the

Step	Reason
	same radius.
3) If I is on circle *C*, then IO = *r*	The radius of the circle (Radius of *C* = *r*).
4) If I is on circle *C'*, then IO' = *r*	The radius of the circle (Radius of *C'*= *r*).
5) I is on the perpendicular bisector of $\overline{OO'}$.	If a point is equidistant from the endpoints of a segment, then it is on its perpendicular bisector.
6) J is on the perpendicular bisector of $\overline{OO'}$.	Same Reasonss as Steps 1-5, for J instead of I.
7) **Therefore, $\overleftrightarrow{IJ}$ is the perpendicular bisector of $\overline{OC}$.**	Two points (I and J) on the perpendicular bisector determine a line ($\overleftrightarrow{IJ}$).

2.6 Two segments $\overline{AB}$ and $\overline{CD}$ intersect at their midpoint M. AB = CD. $\overline{AB}$ and $\overline{CD}$ are **not** perpendicular. L is the symmetric point of A with respect **to the line** $\overleftrightarrow{MC}$. The lines $\overleftrightarrow{LC}$ and $\overleftrightarrow{BD}$ intersect at J.

a. Prove that the angles ∠MCA and ∠MDB are congruent.
b. Prove that triangle JDC is an isosceles triangle.

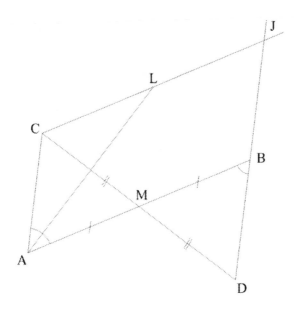

Given:
M is the midpoint of $\overline{AB}$ and $\overline{CD}$; AB = CD.
L = S$_{\overline{MC}}$(A)
$\overleftrightarrow{DB}$ and $\overleftrightarrow{LC}$ intersect at J.
Conclusion:
Prove that angles ∠MCA and ∠MDB are congruent.

Demonstration:

Step	Reason
1) M is the midpoint of $\overline{AB}$,	Given
2) A = S_M(B)	Two points (A and B) are symmetric with respect to a point (M) if the point (M) is the midpoint of the segment ($\overline{AB}$).
3) M is the midpoint of $\overline{CD}$	Given
4) C = S_M(D)	Same Reason as Step 2.
5) $\overleftrightarrow{AC}$ = S_M ($\overleftrightarrow{DB}$)	The symmetric line of a line, with respect to a point, is the line formed by joining the symmetric points of two of its points.
6) $\overleftrightarrow{AC}$ is parallel to $\overleftrightarrow{DB}$	If two lines are symmetric with respect to a point, then they are parallel
7) **Therefore, ∠DCA and ∠CDJ are congruent.**	Alternate interior angles, formed by parallel lines ($\overleftrightarrow{BD}$ is parallel to $\overleftrightarrow{AC}$) and the transversal ($\overleftrightarrow{CD}$), are congruent.
Prove that JDC is an isosceles triangle. 8) L = $S_{\overleftrightarrow{MC}}$ (A)	Given
9) $\overleftrightarrow{MC}$ is the perpendicular bisector of $\overline{AL}$.	Two points (A and L) are symmetric with respect to a line ($\overleftrightarrow{MC}$), if it ($\overleftrightarrow{MC}$) is the perpendicular bisector to the segment ($\overline{AL}$).
10) CL = CA	Any point on the perpendicular bisector of a segment is equidistant from the endpoints of the segment.
11) LCA is an isosceles triangle	If a triangle has two congruent sides, then it is an isosceles triangle.
12) ∠JCD = ∠DCA	In an isosceles triangle (LCA), the perpendicular bisector (CD) of the base ($\overline{AL}$) is the angle bisector of the vertex angle.
13) ∠DCA and ∠CDJ are congruent, Therefore ∠DCA = ∠CDJ.	Refer to Step 7
14) ∠ JCD = ∠CDJ	Transitive property (both equal to ∠ DCA)
15) **Therefore JDC is an isosceles triangle.**	If a triangle has two congruent angles, then it is an isosceles triangle.

2.7 Draw a triangle ABC, and a point I, exterior to the triangle. Draw the symmetric points A', B', C' of A, B, C with respect to I (respectively). Demonstrate that the perpendicular bisectors of $\overline{AB}$ and $\overline{A'B'}$ are parallel.

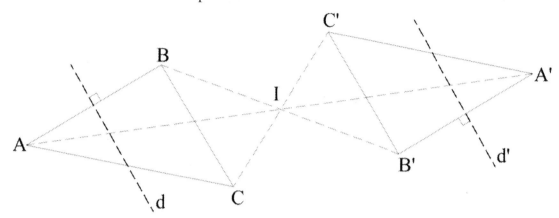

Given:

ABC is a triangle.

I is an exterior point to ABC.

A', B', C' are symmetric points of A, B, C with respect to I.

Conclusion:

Prove that the perpendicular bisectors of $\overline{AB}$ and $\overline{A'B'}$ are parallel.

Demonstration:

Step	Reason
1) A' = S_I(A)	Given
2) B' = S_I(B)	Given
3) Then $\overline{A'B'}$ = S_I($\overline{AB}$)	The symmetric segment of a segment is formed by its symmetric endpoints.
4) $\overline{AB}$ is parallel to $\overline{A'B'}$	If two segments are symmetric with respect to a point then they are parallel.
5) Let $\boldsymbol{d}$ be the perpendicular bisector of $\overline{AB}$, then $\boldsymbol{d}$ is perpendicular to $\overline{AB}$.	Assumption
6) $\boldsymbol{d}$ is also perpendicular to $\overline{A'B'}$	The perpendicular bisector is the line perpendicular at the midpoint of a segment. If two lines are parallel, then any line perpendicular to one is also perpendicular to the other.

Step	Reason
7) Let d' be the perpendicular bisector of $\overline{A'B'}$, then d' is perpendicular to $\overline{A'B'}$.	Same Reason as Steps 1-6.
8) **Therefore d is parallel to d'.**	If two lines (d and d') are perpendicular to a third one $\overline{A'B'}$, then they are parallel.

CHAPTER 3
ANGLES AND PARALLELOGRAMS

Summary of Essential Theorems and Definitions

1) If two parallel lines are cut by a transversal, then:
   a. Corresponding angles are congruent.
   b. Alternate interior angles are congruent.
   c. Alternate exterior angles are congruent.
   d. Angles interior same side of the transversal are supplementary.
   e. Angles exterior same side of the transversal are supplementary.

Conversely

2) If two lines are cut by a transversal forming:
   a. Congruent alternate interior angles,
   b. Congruent alternate exterior angles,
   c. Congruent corresponding angles,
   THEN the lines are parallel.

Please note: In any Demonstration requiring alternate interior angles, or corresponding angles, or alternate exterior angles - you **must** identify the parallel lines and the transversal.

3) In any parallelogram:
   a. The opposite sides are parallel and are congruent,
   b. Opposite angles are congruent,
   c. The diagonals intersect at their midpoint.
   d. Two consecutive angles are supplementary.
   e. The center of the parallelogram is the center of symmetry.

Conversely

4) If a quadrilateral has:
   a. Two pairs of congruent opposite sides, or
   b. Two pairs of parallel opposite sides, or
   c. Two pairs of congruent opposite angles, or
   d. Two diagonals that intersect at their midpoint, or
   e. Two consecutive angles that are supplementary, or
   f. Two opposite sides congruent and parallel at the same time
   THEN it is a parallelogram.

Example

ABCD is a trapezoid with bases $\overline{AB}$ and $\overline{DC}$ such that AB = BC.
Prove that $\overrightarrow{AC}$ is the bisector angle of ∠BCD.

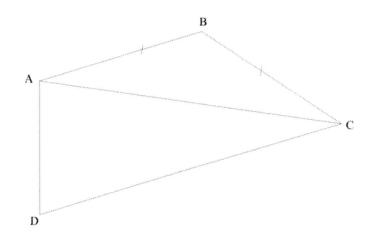

Given:

ABCD is a trapezoid.

$\overline{AB}$ and $\overline{DC}$ are its bases.

AB = BC

Conclusion:

Prove that $\overrightarrow{AC}$ is the bisector of ∠BCD.

Demonstration:

Step	Reason
1) AB = BC 2) Thus ABC is an isosceles triangle 3) And ∠BAC = ∠BCA	Given Any triangle with two congruent sides is an isosceles triangle. An isosceles triangle has two congruent base angles.
4) $\overline{AB}$ is parallel to $\overline{DC}$ 5) ∠BAC = ∠ACD	A trapezoid has two parallel bases. Alternate interior angles, formed by two parallel lines ($\overleftrightarrow{AB}$ and $\overleftrightarrow{DC}$) and a transversal ($\overleftrightarrow{AC}$), are congruent.
6) ∠BCA = ∠ACD 7) **Therefore, $\overrightarrow{AC}$ is the angle bisector of ∠BCD**	Transitive property (both are congruent to ∠BAC). A ray cutting an angle into two congruent angles is the angle bisector.

Problems

3.1 Draw acute angle ∠xEy. Then draw angle ∠EFG = 46° with point F on $\overrightarrow{Ex}$ and point G on $\overrightarrow{Ey}$. The ray $\overrightarrow{FH}$ is the angle bisector of ∠GFx, point H is on $\overrightarrow{Ey}$. Through H, draw $\overleftrightarrow{HI}$ parallel to $\overrightarrow{FG}$. I is a point on $\overrightarrow{Ex}$, and $\overrightarrow{IK}$ is the angle bisector of angle ∠xIH. K is a point on $\overrightarrow{Ey}$.

   a. Calculate the measurements of angles ∠xFG and ∠xFH.

   b. Demonstrate that ∠EFG = ∠ FIH

   c. What is the measurement of ∠xIK?

   d. Demonstrate that $\overleftrightarrow{FH}$ is parallel to $\overrightarrow{IK}$.

3.2 Draw two parallel lines $\overleftrightarrow{xy}$ and $\overleftrightarrow{zt}$. N and Q are two points on $\overleftrightarrow{xy}$; M and P are two points on $\overrightarrow{zt}$ such that $\overrightarrow{NP}$ is the angle bisector of ∠MNQ, and $\overrightarrow{PQ}$ is the angle bisector of ∠NPt.

   a. If ∠NMP = 58°, calculate the value of ∠PNQ and ∠QPT.

   b. Are $\overleftrightarrow{MN}$ And $\overleftrightarrow{PQ}$ parallel lines. Why or why not?

   c. What is the nature of triangles MNP and NPQ?

   d. What value must ∠NMP have for $\overleftrightarrow{MN}$ and $\overleftrightarrow{PQ}$ to be parallel?

3.3 Draw a triangle MAB, and Q is the midpoint of the side $\overline{AB}$. Construct C, the symmetric point of M with respect to Q. Demonstrate that MACB is a parallelogram.

3.4 Draw three points R, S, T, not on the same line. Let U be the midpoint of $\overline{ST}$. The parallel line to $\overline{RS}$, passing through T, and the parallel line to $\overline{RT}$, passing through S, intersect at V. Demonstrate that U is the midpoint of $\overline{RV}$.

3.5 Draw a triangle LMN, and a parallel line to $\overline{MN}$ that cuts $\overline{LM}$ and $\overline{LN}$ at R and S, respectively. Draw I, the midpoint of $\overline{SM}$. K is the symmetric point of point R with respect to I. Demonstrate that K lays on $\overline{MN}$.

3.6 Draw a parallelogram LMNP. The angle bisector of ∠PLM cuts $\overline{PN}$ at K; the angle bisector of ∠MNP cuts $\overline{LM}$ at J.

   a. Demonstrate that LJNK is a parallelogram.

   b. Demonstrate that segments $\overline{LN}$, $\overline{PM}$ and $\overline{KJ}$ are concurrent.

3.7 Draw a right triangle FEG, with right angle E, such that FG = 7cm, and ∠FGE=40°. The perpendicular bisector of $\overline{EG}$ cuts $\overline{FG}$ at O.

   a. Calculate the angles of triangles EOG and EOF.

   b. Demonstrate that O is the midpoint of $\overline{GF}$.

   c. What is the circumcenter of triangle EFG?

Solutions

3.1 Draw acute angle ∠xEy. Then draw angle ∠EFG = 46° with point F on $\overrightarrow{Ex}$ and point G on $\overrightarrow{Ey}$. The ray $\overrightarrow{FH}$ is the angle bisector of ∠GFx, point H is on $\overrightarrow{Ey}$. Through H, draw $\overleftrightarrow{HI}$ parallel to $\overrightarrow{FG}$. I is a point on $\overrightarrow{Ex}$, and $\overrightarrow{IK}$ is the angle bisector of angle ∠xIH. K is a point on $\overrightarrow{Ey}$.

   a. Calculate the measurements of angles ∠xFG and ∠xFH.

b. Demonstrate that ∠EFG = ∠ FIH
c. What is the measurement of ∠xIK?
d. Demonstrate that $\overleftrightarrow{FH}$ is parallel to $\overleftrightarrow{IK}$.

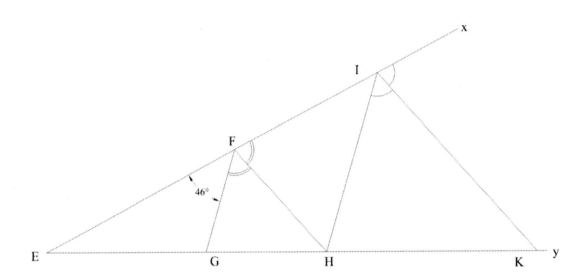

Given:

∠ xEy is an acute angle

∠EFG = 46°

$\overrightarrow{FH}$ is an angle bisector of ∠GFx

$\overleftrightarrow{HI}$ is parallel to FG

$\overrightarrow{IK}$ is the angle bisector of ∠xIH

Conclusion:

Find the measurements of ∠xFG and ∠xFH.

Demonstration:

Step	Reason
1) ∠xFG = ∠xFE - ∠EFG	Subtracting angles
2) ∠xFE = 180°	A straight angle measures 180°.
3) ∠EFG = 46°	Given
4) **Therefore ∠xFG = 180°- 46° = 134°**	Substituting angles by their values.
	Given
5) $\overrightarrow{FH}$ is the angle bisector of ∠GFx	An angle bisector ($\overrightarrow{FH}$) divides an angle (∠xGF) into two congruent angles.
6) ∠xFH = ∠xFG ÷ 2	See Step 4.
	Substituting values.

Step	Reason
7) ∠xFG = 134° 8) **Therefore ∠xFH = 134° ÷ 2 = 67°**	
Demonstrate that ∠EFG = ∠FIH 9) ∠ **EFG** = ∠EIH = **∠FIH** = 46°	If two parallel lines ($\overrightarrow{FG}$ and $\overleftrightarrow{IH}$) are cut by a transversal ($\overrightarrow{EX}$), then the corresponding angles are congruent.
What is the measurement of ∠xIK? 10) xIF = 180° 11) ∠xIH = ∠ xIF - ∠FIH 12) ∠xIH = 180° - 46° = 134° 13) $\overrightarrow{IK}$ is the angle bisector of ∠xIH 14) **Therefore, ∠xIK = 134° ÷ 2 = 67°**	A straight angle measures 180°. Subtracting angles Substituting angles by their values. Given An angle bisector ($\overrightarrow{IK}$) cuts an angle (∠xIH) into two congruent angles.
Demonstrate that $\overleftrightarrow{FH}$ is parallel to $\overleftrightarrow{IK}$. 15) ∠xIK = 67° 16) ∠xFH = 67° 17) Then ∠ xIK = ∠xFH 18) **Therefore, $\overleftrightarrow{IK}$ is parallel to $\overleftrightarrow{FH}$**	See Step 14. See Step 8. Transitive property (both ∠xIK and ∠xFH are equal to 67°). If two lines ($\overleftrightarrow{IK}$ and $\overleftrightarrow{FH}$) cut by a transversal ($\overrightarrow{Ex}$) form congruent corresponding angles, then they are parallel.

3.2 Draw two parallel lines $\overleftrightarrow{xy}$ and $\overleftrightarrow{zt}$. N and Q are two points on $\overleftrightarrow{xy}$; M and P are two points on $\overleftrightarrow{zt}$ such that $\overrightarrow{NP}$ is the angle bisector of ∠MNQ, and $\overrightarrow{PQ}$ is the angle bisector of ∠NPt.

a. If ∠NMP = 58°, calculate the value of ∠PNQ and ∠QPT.
b. Are $\overleftrightarrow{MN}$ and $\overleftrightarrow{PQ}$ parallel lines. Why or why not?
c. What is the nature of triangles MNP and NPQ?
d. What value must ∠NMP have for $\overleftrightarrow{MN}$ and $\overleftrightarrow{PQ}$ to be parallel?

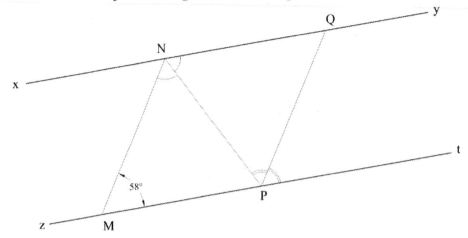

Given:

$\overleftrightarrow{xy}$ is parallel to $\overleftrightarrow{zt}$.

$\overrightarrow{NP}$ is the angle bisector of ∠MNQ.

$\overrightarrow{PQ}$ is the angle bisector of ∠NPt.

∠NMP = 58°

Conclusion:

Find the value of ∠PNQ and ∠QPT.

Demonstration:

Step	Reason
Calculate ∠PNQ. 1) ∠xNM + ∠MNy = ∠xNy = 180°	A straight angle measures 180°.
2) MNy = 180° - ∠xNM	Subtracting the same value (∠ xNM) on both sides of the equation.
3) ∠xNM = ∠NMP = 58°	Alternate-interior angles are congruent ($\overleftrightarrow{xy}$ is parallel to $\overleftrightarrow{zt}$ with transversal $\overleftrightarrow{MN}$).
4) ∠MNy = 180° − 58° = 122°	Substituting ∠ xNM in Step 2 by its value.
5) $\overrightarrow{NP}$ is the angle bisector of ∠MNy	Given
6) ∠PNQ = ∠ MNy ÷ 2	The angle bisector cuts the angle into two congruent angles.
7) **Therefore ∠PNQ = 122° ÷ 2 = 61°**	
Calculate ∠QPt. 8) ∠zPN + ∠NPt = ∠zPt = 180°	A straight angle measures 180°.
9) ∠NPt = 180°- ∠zPN	Subtracting the same value (∠zPN) on both sides of the equation.
10) ∠zPN = ∠PNQ = 61°	Alternate-interior angles are congruent ($\overleftrightarrow{xy}$ is parallel to $\overleftrightarrow{zt}$ with transversal $\overrightarrow{NP}$).

Chapter 3 – Angles and Parallelograms

Step	Reason
11) ∠NPt = 180° - 61° = 119° 12) **Therefore ∠QPt = 119° ÷ 2 = 59.5°**	Substituting ∠zPN in Step 9 by its value. $\overrightarrow{PQ}$ is the angle bisector of ∠NPt
Are $\overleftrightarrow{MN}$ And $\overrightarrow{PQ}$ parallel lines. Why or why not? 13) **$\overleftrightarrow{MN}$ And $\overrightarrow{PQ}$ are not parallel lines.**	The corresponding angles ∠NMP (58°) and ∠QPt (59.5°) are not congruent.
What is the nature of triangles MNP and NPQ? 14) ∠QNP = ∠NPM 15) ∠QNP = ∠MNP 16) ∠NPM = ∠MNP 17) **Therefore MNP is an isosceles triangle.** 18) ∠QPt = ∠NQP 19) ∠QPt = ∠NPQ 20) ∠NPQ = ∠NQP 21) **Therefore NPQ is an isosceles triangle.**	Alternate interior angles are congruent ($\overleftrightarrow{xy}$ is parallel to $\overrightarrow{zt}$ with transversal $\overrightarrow{NP}$). The angle bisector ($\overrightarrow{NP}$) cuts the angle (∠MNQ) into two congruent angles. Transitive property (both ∠NPM and ∠MNP are equal to ∠QNP) A triangle with two congruent angles is an isosceles triangle. Alternate interior angles are congruent ($\overleftrightarrow{xy}$ is parallel to $\overrightarrow{zt}$ with transversal $\overrightarrow{PQ}$). The angle bisector ($\overrightarrow{PQ}$) of an angle (∠NPt), cuts the angle into two congruent angles. Transitive property (both ∠NQP and ∠NPQ are equal to ∠QPt) A triangle with two congruent angles is a isosceles triangle.
What value must ∠NMP have for $\overleftrightarrow{MN}$ and $\overrightarrow{PQ}$ to be parallel? 22) Let x° be the measurement of angle ∠QPt 23) ∠QPt = ∠NPQ = x° 24) ∠MNP = ∠PNQ = x° 25) ∠NPQ = ∠MNP = x°	Assumption The angle bisector ($\overrightarrow{PQ}$) of an angle (∠NPt), cuts the angle into two congruent angles. Alternate interior angles are congruent ($\overleftrightarrow{MN}$ is parallel to $\overrightarrow{PQ}$ with transversal $\overrightarrow{NP}$). The angle bisector ($\overrightarrow{NP}$) of an angle (∠MNQ),

Step	Reason
	cuts the angle into two congruent angles.
26) ∠PNQ = ∠MPN = x°	Alternate interior angles are congruent ($\overleftrightarrow{xy}$ is parallel to $\overleftrightarrow{zt}$ with transversal $\overleftrightarrow{NP}$).
27) ∠QPt + ∠NPQ + ∠MPN = ∠MPt	Adding angles.
28) ∠MPt = 180°	A straight angle measures 180°.
29) x° + x° + x° = 180° 3x = 180° x = 180° ÷ 3 = 60° ∠QPt = x° = 60°	Substituting angles for their values.
30) ∠NMP = ∠QPT	Corresponding angles are congruent ($\overleftrightarrow{MN}$ is parallel to $\overrightarrow{PQ}$ with transversal $\overrightarrow{zt}$).
31) **Therefore ∠NMP should measurement x°, or 60°.**	Transitive property.

3.3 Draw a triangle MAB, and Q is the midpoint of side $\overline{AB}$. Construct C, the symmetric point of M with respect to Q. Demonstrate that MACB is a parallelogram.

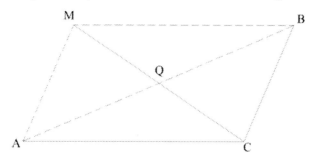

Given:

MAB is a triangle

Q is the midpoint of $\overline{AB}$

$C = S_Q(M)$

Conclusion:

Prove that MACB is a parallelogram.

Demonstration:

Step	Reason
1) Q is the midpoint of $\overline{AB}$.	Given
2) M and C are symmetric points with respect to Q.	Given
3) Q is the midpoint of $\overline{MC}$.	If two points (M and C) are symmetric points with respect to a third point (Q), then the third point (Q) is the midpoint of the segment $\overline{MC}$
4) $\overline{AB}$ and $\overline{MC}$ have the same midpoint Q.	See Step 3.
5) **Therefore, MACB is a parallelogram.**	If the diagonals ($\overline{AB}$ and $\overline{MC}$) of a quadrilateral (MACB) intersect at their midpoint, then the quadrilateral is a parallelogram.

3.4 Draw three points R, S, T, not on the same line. Let U be the midpoint of $\overline{ST}$. The parallel line to $\overline{RS}$, passing through T, and the parallel line to $\overline{RT}$, passing through S, intersect at V. Demonstrate that U is the midpoint of $\overline{RV}$.

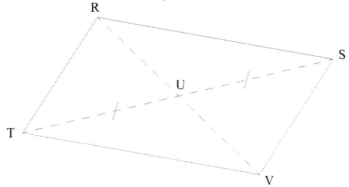

Given:

R, S, and T are three points not on the same line.

U is the midpoint of $\overline{ST}$.

$\overline{SV}$ is a parallel to $\overline{RT}$.

$\overline{RS}$ is a parallel to $\overline{TV}$.

Conclusion:

Prove that U is the midpoint of $\overline{RV}$.

Demonstration:

Step	Reason
Need to prove RSVT is a parallelogram. 1) $\overline{SV}$ is parallel to $\overline{RT}$. 2) $\overline{RS}$ is parallel to $\overline{TV}$. 3) Then RSVT is a parallelogram with diagonals $\overline{RV}$ and $\overline{ST}$.	Given Given A quadrilateral with opposite parallel sides is a parallelogram.
Need to prove U is the midpoint of $\overline{RV}$. 4) $\overline{RV}$ and $\overline{ST}$ are diagonals. 5) U is the midpoint of $\overline{ST}$ 6) **Then U is the midpoint of $\overline{RV}$**	See Step 3. Given In a parallelogram, the diagonals intersect at their midpoint.

3.5 Draw a triangle LMN, and a parallel line to $\overline{MN}$ that cuts $\overline{LM}$ and $\overline{LN}$ at R and S, respectively. Draw I, the midpoint of $\overline{SM}$. K is the symmetric point of point R with respect to I. Demonstrate that K lies on $\overline{MN}$.

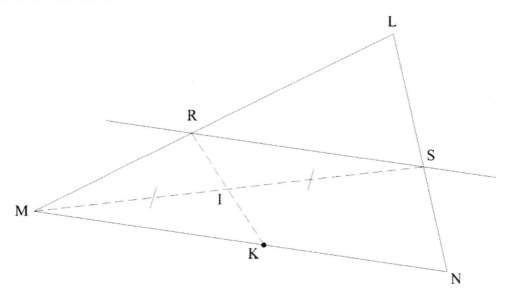

Given:

LMN is a triangle.

$\overleftrightarrow{RS}$ is a parallel line to $\overline{MN}$.

I is the midpoint of $\overline{MN}$.

K is the symmetric point of R with respect to point I

Conclusion:

Prove K lies on $\overline{MN}$.

Demonstration:

Step	Reason
1) I is the midpoint of $\overline{SM}$	Given
2) M = $\mathbf{S}_I$(S)	Two points (M and S) are symmetric with respect to a point (I) if this point (I) is the midpoint of the segment ($\overline{MS}$).
3) K = $\mathbf{S}_I$(R)	Given
4) $\overleftrightarrow{MK} = \mathbf{S}_I(\overleftrightarrow{RS})$	A symmetric line is formed by connecting the symmetric points of two of its points.
5) $\overleftrightarrow{MK}$ is parallel to $\overleftrightarrow{RS}$.	Two symmetric lines with respect to a point are parallel.
6) $\overleftrightarrow{MN}$ is parallel to $\overleftrightarrow{RS}$	Given
7) $\overleftrightarrow{MK}$ is parallel to $\overleftrightarrow{RS}$	See Step 5.
8) **Therefore, $\overleftrightarrow{MN}$ and $\overleftrightarrow{MK}$ are on the same line – and K lies on $\overleftrightarrow{MN}$.**	From a point (M) outside a line, one and only one parallel line can be drawn to this line.

3.6 Draw a parallelogram LMNP. The angle bisector of ∠PLM cuts $\overline{PN}$ at K; the angle bisector of ∠MNP cuts $\overline{LM}$ at J.

   a.   Demonstrate that LJNK is a parallelogram.

   b.   Demonstrate that segments $\overline{LN}$, $\overline{PM}$ and $\overline{KJ}$ are concurrent.

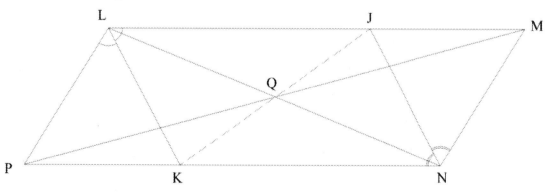

Given:

LMNP is a parallelogram.

$\overrightarrow{LK}$ is the angle bisector of ∠PLM.

$\overrightarrow{PN}$ is the angle bisector of ∠MNP.

Conclusion:

Prove LJNK is a parallelogram.

Demonstration:

Step	Reason
1) $\overline{LM}$ is parallel to $\overline{PN}$	The opposite sides of a parallelogram are parallel.
2) $\overline{LJ}$ is parallel to $\overline{KN}$	Segments included in the opposite sides of a parallelogram are parallel.
3) ∠PLM = ∠PNM	In a parallelogram opposite angles are congruent.
4) $\overrightarrow{LK}$ is the angle bisector of ∠PLM	Given
5) ∠KLM = ∠PLM ÷ 2	The angle bisector cuts an angle into two congruent angles.
6) ∠KLM = ∠PKL	Alternate interior angles, formed by two parallel lines ($\overleftrightarrow{LM}$ and $\overleftrightarrow{PN}$) and a transversal ($\overleftrightarrow{LK}$), are congruent.
7) Therefore ∠PKL = ∠PLM ÷ 2	Transitive property
8) ∠PNM = ∠PLM	In a parallelogram, opposite angles are congruent.
9) $\overrightarrow{NJ}$ is the angle bisector of ∠PNM	Given
10) ∠PNJ = ∠PNM ÷ 2	The angle bisector cuts an angle into two congruent angles.
11) ∠PNM = ∠PLM	In a parallelogram opposite angles are congruent.
12) ∠PNJ = ∠PLM ÷ 2	Substitution (∠PNM in Step 10 by ∠PLM).
13) ∠PKL = ∠PLM ÷ 2	See Step 7.
14) Therefore ∠PKL = ∠PNJ	Transitive property
15) $\overleftrightarrow{LK}$ is parallel to $\overleftrightarrow{JN}$	If two lines, with a transversal, form two congruent corresponding angles, then they are parallel
16) $\overline{LJ}$ is parallel to $\overline{KN}$	See Step 2.
17) **Therefore LJNK is a parallelogram.**	A quadrilateral with a pair of opposite parallel sides is a parallelogram.
$\overline{LN}, \overline{PM}$ and $\overline{KJ}$ are concurrent. 18) Q is the midpoint of $\overline{LN}$ and $\overline{KJ}$.	In a parallelogram (LJNK) the diagonals

Step	Reason
19) Q is the midpoint of $\overline{LN}$ and $\overline{PM}$.	($\overline{LN}$ and $\overline{KJ}$) intersect at their midpoint. In a parallelogram (LMNP) the diagonals ($\overline{LN}$ and $\overline{PM}$) intersect at their midpoint.
20) Midpoint Q is common to $\overline{LN}$, $\overline{KJ}$ and $\overline{MP}$.	Q is located on the three segments.
21) **Therefore $\overline{LN}$, $\overline{PM}$ and $\overline{KJ}$ are concurrent.**	Concurrent segments have the same intersecting point.

3.7 Draw a right triangle FEG, with right angle E, such that FG = 7cm, and ∠FGE=40°. The perpendicular bisector of $\overline{EG}$ cuts $\overline{FG}$ at O.

   a.   Calculate the angles of triangles EOG and EOF.

   b.   Demonstrate that O is the midpoint of $\overline{GF}$.

   c.   What is the circumcenter of triangle EFG?

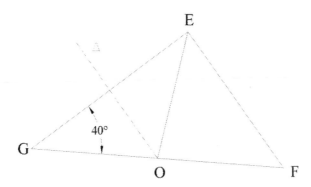

Given:

FEG is a right triangle.

∠GEF a right angle.

FG = 7cm

∠FGE = 40°

Δ is the perpendicular bisector of $\overline{EG}$, and intersects $\overline{GF}$ at O

Conclusion:

Calculate the angles of triangle EOG.

Demonstration:

Step	Reason
1) In the triangle EOG, ∠FGE = 40°.	Given
2) Δ is the perpendicular bisector of $\overline{EG}$.	Given
3) OG = OE	Any point on the perpendicular bisector of a segment is equidistant from the endpoints of this

Step	Reason
4) Triangle GOE is an isosceles triangle. 5) **Therefore ∠EGO = ∠GEO = 40°.** 6) ∠EOG =180° - (∠OGO + ∠GEO) 7) ∠EOG = 180° - (40° + 40°). **Therefore ∠EOG = 100°.**	segment. A triangle with two congruent sides is isosceles. The two base angles of an isosceles triangle are congruent. (∠EGO = ∠GEO = same angle) The sum of the angles of a triangle (EOG) is equal to 180°. Substituting angles by their value.
Calculate the angles of triangle EOF. 8) ∠EFG = 90° - ∠EGF 9) ∠EFG = 90° - 40° = 50°. 10) ∠FEO = ∠FEG - ∠GEO 11) **Therefore ∠FEO = 90° - 40°** **∠FEO = 50°** 12) ∠FOE = 180° – (∠FEO + ∠OFE) 13) **Therefore ∠FOE = 180° – 100°** **∠FOE = 80°**	In a right triangle (FEG) the acute angles are complementary. Substituting the angles by their values. Subtracting angles. Substituting the angles by their values. The sum of the angles of a triangle (FOE) is equal to 180°. Substituting angles by their values.
Demonstrate that O is the midpoint of $\overline{GF}$. 14) OG = OE 15) ∠FEO = ∠OFE = 50° 16) EOF is an isosceles triangle 17) OE = OF 18) OG = OF 19) O is on $\overline{GF}$ 20) **Therefore O is the midpoint of segment $\overline{GF}$.**	See Step 3. See Steps 8 and 9. If a triangle has two congruent angles, then it is an isosceles triangle. In an isosceles triangle, the two legs are congruent. Transitive property (both congruent to $\overline{OE}$) Given The midpoint is the point halfway between the endpoints of a segment.
What is the circumcenter of triangle EFG? 21) OE = OF = OG 22) O is the center of a circle passing through F, E, and G. 23) **Therefore O is the circumcenter of triangle FEG.**	See Steps 17 and 18. A circle is the set of points at the same distance from the center. A circumscribed circle passes through the three vertices of the triangle. (F, E, and G).

CHAPTER 4
SPECIAL TRIANGLES

Summary of Essential Theorems and Definitions

1) Isosceles Triangles

If a triangle is an isosceles triangle, then:

    a. Its base angles are congruent.

    b. Its legs are congruent.

    c. The median from the vertex angle is the same as the altitude, the angle bisector, and the perpendicular bisector.

Conversely, if a triangle has:

    d. Congruent base angles, or

    e. Congruent legs, or

    f. Its median from the vertex angle is the same as its altitude, angle bisector, or perpendicular bisector,

THEN it is an isosceles triangle.

2) Right Triangles

If a triangle is a right triangle, then:

    a. The acute angles are complementary.

Conversely, if a triangle has:

    b. Complementary angles,

THEN it is a right triangle.

3) Equilateral Triangles

If a triangle is an equilateral triangle, then:

    a. Each angle measures 60°.

    b. **Any** altitude is the same as the median, the angle bisector and the perpendicular bisector.

Conversely, if a triangle has:

    c. Three 60° angles, or

    d. **All** altitudes that are the same as the median, the angle bisector or the perpendicular bisector.

THEN it is an equilateral triangle.

Example

Draw a circle with center O and the radius OM = 3cm. $\overline{MP}$ is the diameter. Draw ∠NOP = 50°.

1) Find the measurement of angle ∠NOM.
2) What is the nature of triangle NOM?
3) Find the measurement of angle ∠NMP.

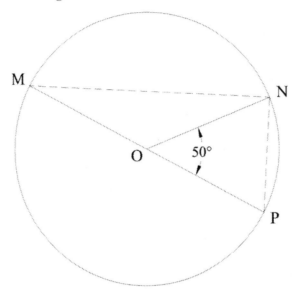

Given:

A circle with center O, radius OM = 3cm, and diameter *MP*

∠NOP = 50°

Conclusion:

Find the measurement of angle ∠NOM.

Demonstration:

Step	Reason
1) $\overline{MP}$ is the diameter	Given
2) ∠MOP = 180°	A straight angle measures 180°.
3) ∠NOM = ∠MOP - ∠NOP	Subtracting angles
4) **Therefore ∠ NOM = 180° - 50°** **∠ NOM = 130°.**	Substituting angles with their values.
What is the nature of triangle NOM? 5) OM = ON 6) **Therefore NOM is an isosceles triangle.** 7) And ∠OMN = ∠ONM	The radii of the same circle are congruent. A triangle with two congruent legs is an isosceles triangle. In an isosceles triangle the two base angles are congruent.

Step	Reason
Calculate ∠NMP 8) ∠NMP = (180° - ∠NOM) ÷ 2	The sum of the angles in a triangle is 180° And, the base angles of an isosceles triangle are congruent.
9) **∠NMP = 50° ÷ 2 = 25°**	Substituting angles by their values.

Problems

4.1 Draw two circles **C** and **C'** with the same center O, and respective radius r and r'. Let M and N be two points on **C**, and L and P be two points on **C'**, such that $\overline{LN}$ and $\overline{MP}$ intersect at O. ∠LOP = 46°.

  a. What is the exact nature of triangle LOP?

  b. Find the measurement of ∠OLP, and ∠MNO.

  c. Prove that $\overleftrightarrow{LP}$ and $\overleftrightarrow{MN}$ are parallel.

4.2 Draw a circle **C** with center O, and a radius of 3cm. Add the radius $\overline{OL}$. Place a point M on this circle such that ∠OLM = 65°. Place the point N such that ∠OMN = 50°, and M is between L and N.

  a. Find the measurement of ∠LOM.

  b. What is the nature of the quadrilateral LMNO?

4.3 On the following drawing ∠DAC = 88°, ∠ADC = 52°, ∠ABC = 76°, and ∠ACB = ¾ of ∠DAC.

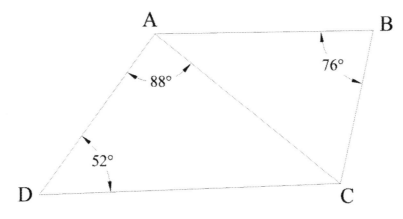

  a. Find the measurement of the missing angles?

  b. Are lines $\overleftrightarrow{AB}$ and $\overleftrightarrow{DC}$ parallel?

4.4 Draw an isosceles triangle BAC, with A as the vertex angle, such that AB = 5cm and ∠BAC = 48°. The angle bisectors of ∠ABC and ∠ACB intersect at O.

  a. Find the measurements of the angles of triangle BOC.

  b. If Q is the midpoint of $\overline{BC}$, demonstrate that A, O, and Q are on the same line.

4.5 Draw triangle ABE such that AE = 4cm, ∠BAE = 118° and ∠AEB = 26°.

   a. Draw the triangle BCD such that BC = 3cm, ∠CBD = 48° and $\overline{ED}$ is parallel to $\overline{AB}$. ∠EBD = 2 × ∠BDE.

   b. Are the points A, B, and C on the same line?

4.6 Draw two parallel lines $\overleftrightarrow{xy}$ and $\overleftrightarrow{zt}$. Place the point L on $\overleftrightarrow{xy}$. Draw ∠xLM = ∠66° so that M is on $\overleftrightarrow{zt}$. Draw the circle with center L and radius $\overline{LM}$ so that it cuts $\overleftrightarrow{xy}$ at P, and $\overleftrightarrow{zt}$ at M and N. Draw the perpendicular line to $\overleftrightarrow{xy}$ passing through P, so that it cuts $\overleftrightarrow{zt}$ at Q. By Q draw ∠tQR = 57° with R on $\overleftrightarrow{xy}$.

   a. Find the measurement of angle ∠PNQ.

   b. Demonstrate that PR = NQ.

Solutions

4.1 Draw two circles C and C' with the same center O, and respective radius r and r'. Let M and N be two points on C, and L and P be two points on C', such that $\overline{LN}$ and $\overline{MP}$ intersect at O. ∠LOP = 46°.

   a. What is the exact nature of triangle LOP?

   b. Find the measurement of ∠OLP, and ∠MNO.

   c. Prove that $\overleftrightarrow{LP}$ and $\overleftrightarrow{MN}$ are parallel.

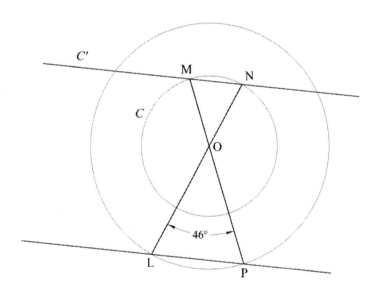

Given:

C and C' are circles with the same center O, and radii r and r'.

The lines $\overleftrightarrow{LN}$ and $\overleftrightarrow{PM}$ intersect at O.

∠LOP = 46°

Conclusion:

Find the exact nature of triangle LOP.

Demonstration:

Step	Reason
1) OL = OP	The radii of a circle have the same measurement (C' $\Rightarrow$ r')
2) **Therefore, triangle LOP is an isosceles triangle.**	A triangle with two congruent sides is an isosceles triangle.
3) And $\angle$OLP = $\angle$OPL	In an isosceles triangle, the base angles are congruent.
Find the measurement of $\angle$OLP.	
4) $\angle$OLP = (180° - $\angle$LOP) ÷ 2	The sum of the angles of a triangle is 180°, and the base angles of an isosceles triangle are congruent.
5) $\angle$OLP = (180° - 46°) ÷ 2	Substituting the angles by their values.
6) $\angle$OLP = 134° ÷ 2 = 67°	
7) **Therefore $\angle$OLP = 67°**	
Find the measurement of $\angle$MNO.	
8) $\angle$MON = $\angle$LOP = 46°.	Vertical angles are congruent.
9) MO = NO = r	The radii of a circle are congruent (C $\Rightarrow$ r)
10) The triangle MON is isosceles.	A triangle with two congruent sides is isosceles.
11) Then $\angle$MNO = $\angle$OMN	The base angles of an isosceles triangle are congruent.
12) $\angle$MNO = (180° - 46°) ÷ 2	The sum of the angles of a triangle is 180°, and the base angles of an isosceles triangle are congruent.
13) $\angle$MNO = 134° ÷ 2 = 67°	Substituting the angles by their values.
14) **Therefore $\angle$MNO = 67°**	
Prove that $\overleftrightarrow{LP}$ and $\overleftrightarrow{MN}$ are parallel.	
15) $\angle$OLP and $\angle$MNO are alternate-interior angles	Angles formed by two lines ($\overleftrightarrow{LP}$ and $\overleftrightarrow{MN}$) and a transversal ($\overleftrightarrow{LN}$).
16) **Therefore $\overleftrightarrow{LP}$ and $\overleftrightarrow{MN}$ are parallel**	If two lines with a transversal form congruent alternate-interior angles, then they are parallel.

4.2 Draw a circle **C** with center O, and a radius of 3cm. Add the radius $\overline{OL}$. Place a point M on this circle such that $\angle$OLM = 65°. Place the point N such that $\angle$OMN = 50°, and M is between L and N.

   a. Find the measurement of $\angle$LOM.

   b. What is the nature of the quadrilateral LMNO?

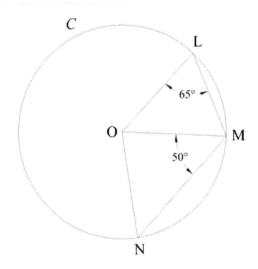

Given:

C is a circle with center O.

$\angle$OLM = 65°

$\angle$OMN = 50°

Conclusion:

Find the measurement of $\angle$LOM.

Demonstration:

Step	Reason
1) OL = OM	The radii of a circle have the same measurement.
2) LOM is an isosceles triangle.	A triangle with two congruent sides is an isosceles triangle.
3) $\angle$OLM = $\angle$OML	In an isosceles triangle, the base angles are congruent.
4) $\angle$LOM = 180° - ($\angle$OLM + $\angle$OML)	The sum of the angles of a triangle is 180°.
5) $\angle$OLM = 65°	Given
6) $\angle$LOM = 180° - (65° + 65°)	Substituting the angles by their values.
7) **Therefore $\angle$LOM = 180° - 130° = 50°**	
Measurement of $\angle$NOM 8) ON = OM	The radii of a circle have the same measurement.
9) NOM is an isosceles triangle.	A triangle with two congruent sides is an isosceles triangle.
10) $\angle$OMN = $\angle$ONM	In an isosceles triangle, the base angles are congruent.
11) $\angle$NOM = 180° - ($\angle$OMN + $\angle$ONM)	The sum of the angles of a triangle is 180°.
12) $\angle$OMN = 50°	Given
13) $\angle$NOM = 180° - (50° + 50°)	Substituting the angles by their values.

Step	Reason
Therefore ∠NOM = 180° - 100° = 80°	

The nature of the quadrilateral LMNO?	
14) ∠LON = ∠LOM + ∠ MON	Adding angles.
15) ∠LON = 50° + 80° = 130°	Substituting angles by their values.
16) ∠LOM = 50°	See Step 7.
17) ∠OMN = 50°	Given
18) ∠LOM and ∠OMN are alternate interior angles.	Angles formed by two segments ($\overline{OL}$ and $\overline{NM}$) and a transversal ($\overline{OM}$).
19) $\overline{OL}$ and $\overline{NM}$ are parallel	If two segments ($\overline{OL}$ and $\overline{NM}$) form, with a transversal ($\overline{OM}$), congruent alternate interior angles, then they are parallel.
20) ∠LON = 130°	See Step 15.
21) ∠LMN = ∠LMO + ∠OMN	Adding angles.
22) ∠LMN = 65° + 50° = 110°	Substituting angles by their value.
23) **Then LMNO is a trapezoid**	If a quadrilateral (LMNO) has two opposite parallel sides ($\overline{OL}$ and $\overline{MN}$) and opposite angles (∠LMN and ∠LON), of different value then it is a trapezoid.

4.3 On the following drawing ∠DAC = 88°, ∠ADC = 52°, ∠ABC = 76°, and∠ACB = ¾ of ∠DAC.

  a. Find the measurement of the missing angles?

  b. Are lines $\overleftrightarrow{AB}$ and $\overleftrightarrow{DC}$ parallel?

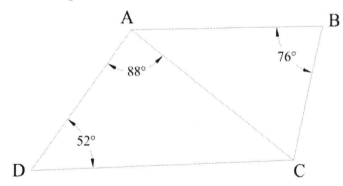

Given:

ABCD is a quadrilateral.

∠DAC = 88°, ∠ADC = 52°, and ∠ABC = 76°

∠ACB = ¾ of ∠DAC

Conclusion:

Find the measures of the missing angles.

Demonstration:

Step	Reason
The value of ∠ACD. 1) ∠ACD + ∠ADC + ∠DAC = 180° 2) ∠ACD + 52° + 88° = 180° 3) ∠ACD + 140° = 180° 4) **Therefore ∠ACD = 180° - 140° = 40°**	The sum of the angles of a triangle is 180°. Substituting angles by their values. Solving the equation for ∠ACD.
The value of ∠ACB. 5) ∠ACB = ¾ ∠ DAC 6) ∠DAC = 88° 7) **Therefore ∠ACB = ¾ × 88° = 66°**	Given Given Solving the equation for ∠ACB.
The value of ∠BAC. 8) ∠BAC = 180° - (∠ACB + ∠ABC) 9) ∠BAC = 180° - (66° + 76°) 10) **Therefore ∠BAC = 38°**	The sum of the angles of a triangle is 180°. Substituting angles by their value. Solving the equation for ∠BAC.
Are lines $\overleftrightarrow{AB}$ and $\overleftrightarrow{DC}$ parallel? 11) ∠BAC and ∠ACD are alternate interior angles. 12) ∠BAC = 38° 13) ∠ACD = 40° 14) **Therefore $\overleftrightarrow{AB}$ and $\overleftrightarrow{DC}$ are not parallel.**	Angles formed by two segments ($\overleftrightarrow{AB}$ and $\overleftrightarrow{DC}$) and a transversal ($\overleftrightarrow{AC}$). See Step 10. See Step 4. For lines to be parallel, their alternate-interior angles should be congruent.

4.4 Draw an isosceles triangle BAC, with A as the vertex angle, such that AB = 5cm and ∠BAC = 48°. The angle bisectors of ∠ABC and ∠ACB intersect at O.

   c. Find the measurements of the angles of triangle BOC.

   d. If Q is the midpoint of $\overline{BC}$, demonstrate that A, O, and Q are on the same line.

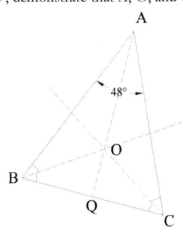

Given:

ABC is an isosceles triangle.

$\angle$BAC is the vertex angle.

$\angle$BAC = 48°

$\overrightarrow{BO}$ and $\overrightarrow{CO}$ are the angle bisectors of $\angle$ABC and $\angle$ACB (respectively).

Conclusion:

Find the measurements of the angles of triangle BOC.

Demonstration:

Step	Reason
1) $\angle$ABC = $\angle$ACB	The base angles of an isosceles triangle (ABC) are congruent.
2) $\angle$ACB = (180° − $\angle$BAC) ÷ 2	The sum of the angles of a triangle is 180°, and Step 1.
3) $\angle$ABC =$\angle$ACB = (180° − 48°) ÷ 2 $\angle$ABC = 132° ÷ 2 = 66°	Substituting $\angle$BAC by its value.
4) **$\angle$OBC = 66° ÷ 2 = 33°**	The bisector angle ($\overrightarrow{BO}$) cuts an angle into two congruent angles.
5) $\angle$ACB=132° ÷ 2 = 66°	Same as Steps 1-3.
6) **$\angle$OCB = 66° ÷2 = 33°.**	The bisector angle ($\overrightarrow{CO}$) cuts an angle into two congruent angles.
7) $\angle$BOC = 180° − ($\angle$OBC + $\angle$OCB)	The sum of the angles of a triangle is 180°.
8) $\angle$BOC = 180° − (33° + 33°)	Substituting the angles by their value
9) **$\angle$BOC = 180° − 66° = 114°**	
Points A, O, and Q are collinear. 10) The angle bisectors of $\angle$ABC and $\angle$ACB intersect at O.	Given
11) O is the incenter of triangle ABC.	The three angle bisectors of a triangle intersect at the incenter (two suffice).
12) $\overrightarrow{AO}$ is the third angle bisector of the triangle.	Three angle bisectors of a triangle intersect at the incenter.
13) In triangle ABC, $\overrightarrow{AQ}$ is the median from the vertex angle.	The median of a triangle is the ray joining the vertex to the midpoint of the opposite side.
14) $\overrightarrow{AQ}$ is also the angle bisector of the vertex angle.	In an isosceles triangle the median from the vertex angle is at the same time its angle bisector.
15) $\overrightarrow{AQ}$ and $\overrightarrow{AO}$ are on the same angle	The points are on the same line ($\overrightarrow{AQ}$ or $\overrightarrow{AO}$)

Step	Reason
bisector. 16) **Therefore A, O and Q are collinear.**	

4.5 Draw triangle ABE such that AE = 4cm, ∠BAE = 118° and ∠AEB = 26°.

   a. Draw the triangle BCD such that BC = 3cm, ∠CBD = 48° and $\overline{ED}$ is parallel to $\overline{AB}$.
      ∠EBD = 2 × ∠BDE.

   b. Are the points A, B, and C on the same line?

<u>Study Drawing:</u>

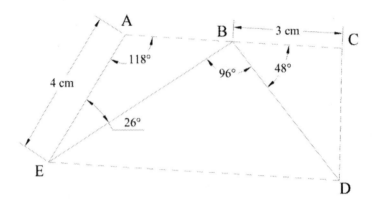

<u>Finding how to draw the figure:</u>

The problem is where to place point D.

Step	Reason
1) D is located on the ray $\overrightarrow{ED}$ parallel to $\overleftrightarrow{AB}$.	Given
2) ∠ABE = 180° - (118° + 26°)	The sum of the angles of a triangle is 180°.
3) ∠ABE = 180° - 144° = 36°	Reducing the equation.
4) ∠BED = ∠ABE = 36°	Alternate interior angles; formed by two parallel lines ($\overleftrightarrow{AB}$ and $\overleftrightarrow{ED}$), and a transversal ($\overleftrightarrow{EB}$) are congruent.
Value of ∠EBD. 5) Let x° be the value of ∠BDE 6) ∠EBD = 2 × ∠BDE 7) ∠BDE + ∠DEB + ∠EBD = 180°	Assumption Given The sum of the angles of a triangle is 180°.

Step	Reason
8) Or: $x° + 36° + 2x° = 180°$	Substituting the angles by their values.
9) $3x° = 180° - 36° = 144°$ $x° = 144° \div 3 = 48°$	Solving the equation.
10) **Therefore $\angle EBD = 2x° = 96°$**	

Explanation of the figure:

Draw $AE = 4cm$; $\angle EAB = 118°$, and $\angle AEB = 26°$

Draw $\angle ABE = 36°$ and $\overrightarrow{Ex}$ parallel to $\overleftrightarrow{AB}$

Draw $\angle EBD = 96°$ (D on $\overrightarrow{ED}$)

Draw $\angle DBC = 48°$ with $BC = 3cm$.

Draw $\overline{DC}$.

Exact drawing:

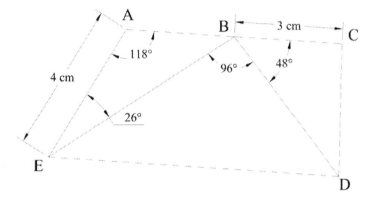

Given:

All of the above.

Conclusion:

Prove that the points A, B, and C are on the same line.

Demonstration:

Step	Reason
1) $\angle ABC = \angle ABE + \angle EBD + \angle DBC$	Adding the angles.
2) $\angle ABC = 36° + 96° + 48° = 180°$	Substituting the angles by their values.
3) $\angle ABC$ is a straight angle	An angle measuring $180°$ is a straight angle.
4) **Therefore A, B, and C are on the same line (collinear).**	

4.6 Draw two parallel lines $\overleftrightarrow{xy}$ and $\overleftrightarrow{zt}$. Place the point L on $\overleftrightarrow{xy}$. Draw $\angle xLM = \angle 66°$ so that M is on $\overleftrightarrow{zt}$. Draw the circle with center L and radius $\overline{LM}$ so that it cuts $\overleftrightarrow{xy}$ at P, and $\overleftrightarrow{zt}$ at M and N. Draw the perpendicular line to $\overleftrightarrow{xy}$ passing through P, so that it cuts $\overleftrightarrow{zt}$ at Q. By Q draw $\angle tQR = 57°$ with R on $\overleftrightarrow{xy}$.

c. Find the measurement of angle $\angle PNQ$.

d. Demonstrate that PR = NQ.

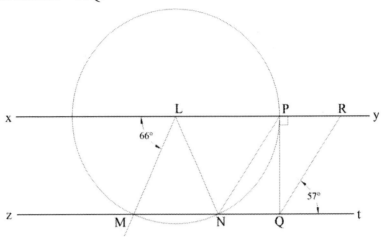

Given:

$\overleftrightarrow{xy}$ is parallel to $\overleftrightarrow{zt}$.

L is a point on $\overleftrightarrow{xy}$.

The circle, with center L, has a radius $\overline{LM}$, and cuts $\overleftrightarrow{xy}$ at P and $\overleftrightarrow{zt}$ at N.

$\overleftrightarrow{PQ}$ is perpendicular to $\overleftrightarrow{xy}$.

$\angle tQR = 57°$

Conclusion:

Find the measurement of $\angle PNQ$.

Demonstration:

Step	Reason
1) $\angle LMN = \angle xLM$	Alternate interior angles formed by two parallel lines ($\overleftrightarrow{xy}$ is parallel to $\overleftrightarrow{zt}$), and a transversal ($\overline{LM}$) are congruent.
2) $\angle xLM = 66°$	Given
3) $\angle LMN = 66°$	Transitive property.
4) LM = LN	Radii of the same circle are congruent.
5) LMN is an isosceles triangle.	A triangle with two congruent angles is an isosceles triangle.
6) $\angle LNM = \angle LMN = 66°$	Base angles of an isosceles triangle are congruent.
7) $\angle LNM = \angle NLP$	Alternate interior angles formed by two parallel lines ($\overleftrightarrow{xy}$ is parallel to $\overleftrightarrow{MN}$) with a transversal ($\overline{LN}$) are congruent.
8) $\angle LNM = 66°$	See Step 6.
9) $\angle NLP = 66°$	Transitive property.
10) LN = LP	The radii of the same circle are congruent.

Step	Reason
11) LNP is an isosceles triangle	A triangle with two congruent sides is isosceles.
12) $\angle$ LNP = $\angle$LPN	Base angles of an isosceles triangle are congruent.
13) $\angle$LNP = $(180° - 66°) \div 2$	The sum of the angles of a triangle is 180°, and Step 9.
14) $\angle$LNP = $\angle$LPN = $114° \div 2 = 57°$	Reducing the equation.
15) $\angle$PNQ = $\angle$LPN	Alternate interiors angles formed by two parallel lines ($\overleftrightarrow{xy}$ and $\overleftrightarrow{zt}$) with transversal $\overleftrightarrow{NP}$ are congruent.
16) $\angle$LNP = $57°$	See Step 14.
17) **Therefore $\angle$PNQ = $\angle$LPN = 57°**	Transitive property
Demonstrate PR = NQ.	
18) $\angle$PNQ = $57°$	See Step 17.
19) $\angle$tQR = $57°$	Given
20) $\angle$PNQ = $\angle$tQR	Transitive property (both equal to 57°)
21) Line $\overleftrightarrow{PN}$ is parallel to line $\overleftrightarrow{RQ}$	If two lines and a transversal form congruent corresponding angles, then they are parallel.
22) $\overleftrightarrow{xy}$ is parallel to $\overleftrightarrow{zt}$.	Given
23) $\overleftrightarrow{xy}$ is also parallel to $\overleftrightarrow{NQ}$.	Lines included in parallel lines are parallel (in $\overleftrightarrow{xy}$ and $\overleftrightarrow{zt}$ respectively).
24) NPRQ is a parallelogram	A quadrilateral with two pairs of opposite parallel sides is a parallelogram.
25) **Therefore PR = NQ**	In a parallelogram opposite sides are congruent.

CHAPTER 5
SPECIAL QUADRILATERALS

Summary of Essential Theorems and Definitions

1) **Rectangle** – All rectangles have:
   a. All the properties of the parallelogram,
   b. Four right angles,
   c. Diagonals with the same measurement, and
   d. Two axes of symmetry (the perpendicular bisectors of the sides).

   Conversely, if a quadrilateral has:
   e. Two opposite sides at the same time congruent and parallel, and a right angle, or
   f. Three right angles, or
   g. One of the properties of the parallelogram and one right angle, or
   h. The diagonals intersecting at their midpoint and having the same measurement, or
   i. One of the properties of the parallelogram and the diagonals of the same measurement,

   THEN it is a rectangle.

2) **Rhombus** – All rhombuses have:
   a. Four congruent sides,
   b. All the properties of the parallelogram, and
   c. Its diagonals intersect perpendicularly at their midpoints.

   Conversely, if a quadrilateral has:
   d. Its four sides of the same measurement, or
   e. One of the properties of the parallelogram and two consecutive sides of the same measurement, or
   f. One of the properties of the parallelogram and the diagonals perpendicular,

   THEN it is a rhombus.

3) **Square** – All squares have:
   a. All the properties of the rectangle, **and** all the properties of the rhombus,
   b. Diagonals that are perpendicular, intersect at their midpoints, and have the same measurements.

   Conversely, if a quadrilateral has:
   c. The properties of a rectangle **and** a rhombus, or
   d. Diagonals that are perpendicular, intersect at their midpoints, and have the same measurements,

   THEN it is a square.

4) **Trapezoid**

   All isosceles trapezoids have:
   a. Non-parallel sides that are congruent.
   b. A trapezoid is isosceles when its axis of symmetry is the perpendicular bisector of its bases.

Example

ABCD is a rhombus; O is its center of symmetry. The perpendicular line to $\overleftrightarrow{AC}$ passing through A cuts the perpendicular line to $\overleftrightarrow{BD}$ passing through B at Q.

What is the nature of quadrilateral BQAO?

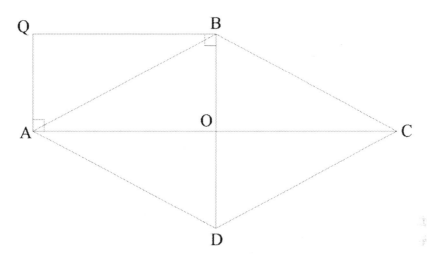

Given:

ABCD is a rhombus.

O is its center of symmetry.

$\overleftrightarrow{QA}$ is a perpendicular line to $\overleftrightarrow{AC}$.

$\overleftrightarrow{QB}$ is a perpendicular line to $\overleftrightarrow{DB}$.

Conclusion:

Find the exact nature of BQAO.

Demonstration:

Step	Reason
1) $\overline{AC}$ and $\overline{DB}$ are the diagonals of the rhombus.	Given
2) O is the center of symmetry.	Given
3) $\overline{AC}$ and $\overline{DB}$ intersect at O	The diagonals of a rhombus intersect at the center of symmetry.
4) $\overline{AC}$ and $\overline{DB}$ are perpendicular	In a rhombus the diagonals are perpendicular.
5) **∠AOB = 90°**	Perpendicular lines form a 90° angle.
6) $\overline{QA}$ is perpendicular to $\overline{OA}$	Given
7) ∠OAQ = 90°	Angle formed by perpendicular lines.
8) $\overline{QB}$ is perpendicular to $\overline{OB}$	Given
9) ∠OBQ = 90°	Angle formed by perpendicular lines (same as Step 7).
10) **Therefore, the quadrilateral BQAO is a rectangle.**	A quadrilateral with three right angles is a rectangle (∠OAB, ∠OAQ and ∠OBQ).

Problems

5.1 LIM is a right triangle where ∠LIM is the right angle. Draw the parallelograms ILMD and IMDE.

    a. What is the exact nature of IMDE?

    b. Draw the parallelogram LMEF.

    c. What is its exact nature?

5.2 Draw the square LMNP such that MP = 6cm and O is its center of symmetry.

    a. Draw the parallelogram LOME. What is its exact nature?

    b. Draw the parallelogram LOPF. What is its exact nature?

    c. Demonstrate that E, L, and F are on the same line

    d. Demonstrate that point L is the midpoint of segment $\overline{EF}$.

    e. Calculate the perimeter of EMPF.

5.3 Draw a ¼ circle $\overparen{QL}$ with radii $\overline{OQ}$ and $\overline{OL}$, where OL = 4cm. M is a point on arc $\overparen{QL}$, $\overline{MN}$ is perpendicular to $\overline{OL}$, $\overline{MP}$ is perpendicular to $\overline{OQ}$. Calculate PN.

5.4 Draw a parallelogram ABCD such that ∠ABC = 78°. $\overrightarrow{DI}$ is the angle bisector of ∠ADC, and $\overrightarrow{AJ}$ the angle bisector of ∠DAB. $\overrightarrow{DI}$ and $\overrightarrow{AJ}$ intersect at point M. AB = 7cm and BC = 4.5cm E is a point on $\overleftrightarrow{AB}$ exterior to ABCD such that AE = 2cm.

    a. Find the measurement of angles ∠ADC, ∠DAE, and ∠DAB.

    b. Find the measurement of angles ∠MDA and ∠DAM. Deduct the nature of triangle DMA and triangle AMI.

    c. Draw the angle bisector of ∠BCD cutting $\overrightarrow{DM}$ at N, and the angle bisector of ∠ABC cutting $\overleftrightarrow{CN}$ at P and $\overleftrightarrow{AM}$ at Q. What is the nature of MNPQ?

5.5 Draw a rhombus ABCD with center I such that AB = 5cm and ∠CAB = 60°. Place E the midpoint of $\overline{BC}$. Place J the symmetric point of point I with respect to E.

    a. Explain the construction.

    b. Demonstrate that IBJC is a rectangle.

    c. Find the values of the angles of triangle IBE.

5.6 Draw a rectangle ABCD with center I such that AB = 6 cm and ∠BAC = 40°. Draw E and F, as the midpoints of $\overline{BC}$ and $\overline{DC}$ (respectively). J is the symmetric point of I with respect to E, and K is the symmetric point of I with respect to F.

    a. Demonstrate that IBJC is a rhombus.

    b. Demonstrate that K, C, and J are aligned, and that C is the midpoint of segment $\overline{KJ}$.

5.7 Draw a rectangle ABCD. The perpendicular lines at the endpoints of the diagonals $\overline{AC}$ and $\overline{BD}$ intersect at E, F, G, and H. Demonstrate that EFGH is a rhombus.

Solutions

5.1 LIM is a right triangle where ∠LIM is the right angle. Draw the parallelograms ILMD and IMDE

    a.   What is the exact nature of IMDE?

    b.   Draw the parallelogram LMEF.

    c.   What is its exact nature?

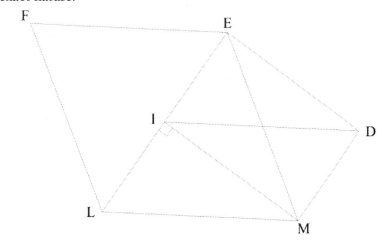

Given:

LIM is a right triangle with ∠LIM as the right angle.

ILMD, IMDE, and LMEF are parallelograms.

Conclusion:

Find the nature of IMDE.

Demonstration:

Step	Reason
1) ∠LIM = 90°.	Given (right angle)
2) $\overleftrightarrow{IL}$ is perpendicular line to $\overleftrightarrow{IM}$.	Legs of a right angle are perpendicular.
3) $\overleftrightarrow{DM}$ is parallel to $\overleftrightarrow{IL}$.	In a parallelogram (ILMD), the opposite sides are parallel.
4) ∠IMD = 90°.	If two lines ($\overleftrightarrow{DM}$ and $\overleftrightarrow{IL}$) are parallel, any perpendicular to one is also perpendicular to the other.
5) IMDE is a parallelogram.	Given
6) IMDE has a right angle.	See Step 4 (∠IMD = 90°)
7) **Therefore IMDE is a rectangle.**	A parallelogram with a right angle is a rectangle.
The exact nature of LMEF.	
8) LM = ID	Opposite sides of a parallelogram are congruent.
9) ID = ME	In a rectangle (IMDE), the diagonals are

Step	Reason
	congruent.
10) LM = ME	Transitive property (both equal to ID).
11) **Therefore LMEF is a rhombus.**	If a parallelogram has two consecutive congruent sides then it is a rhombus.

5.2 Draw the square LMNP such that MP = 6cm and O is its center of symmetry.

    a.   Draw the parallelogram LOME. What is its exact nature?

    b.   Draw the parallelogram LOPF. What is its exact nature?

    c.   Demonstrate that E, L, and F are on the same line

    d.   Demonstrate that point L is the midpoint of segment $\overline{EF}$.

    e.   Calculate the perimeter of EMPF.

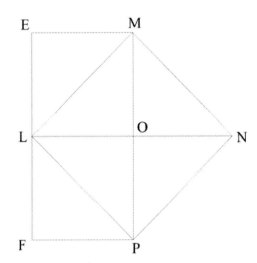

Explanation of the figure:

If $\overline{MP}$ is the diagonal of the square, then is also the angle bisector.

Therefore, ∠LMP = ∠PMN = (∠LMN) ÷ 2

∠LMP = ∠PMN = 90° ÷ 2 = 45° because ∠LMN = 90° (the right angle of the square).

And, ∠LPM = ∠MPN = (∠LPN) ÷ 2

∠LPM = ∠MPN = 90° ÷ 2 = 45° because ∠LPN = 90° (the right angle of the square).

Exact Construction:

We must: a) Draw $\overline{MP}$ = 6cm.

        b) Draw the angles ∠PMN and ∠MPN = 45°.

        c) Draw the square LMNP using the diagonals $\overline{MP}$ and $\overline{LN}$ and their midpoint O.

Given:

All of the above.

O is the center of symmetry.

LOME is a parallelogram.

Conclusion:

Find the exact nature of LOME.

Demonstration:

Step	Reason
1) LOME is a parallelogram.	Given
2) $\overline{LN}$ is perpendicular to $\overline{MP}$.	In a square (LMNP) the diagonals are perpendicular.
3) LOME is a rectangle.	A parallelogram with a right angle is a rectangle.
4) MP = 6cm	Given
5) LN = MP = 6cm	In a square the diagonals are congruent.
6) LO = ON and MO = OP	In a square the diagonals intersect at their midpoint.
7) LO = MO = 3cm	Half the measurement of congruent diagonals.
9) **Therefore LOME is a square.**	A rectangle with two consecutive congruent sides is a square.
Nature of LOPF 10) **LOPF is a square.**	Same proof as for LOME.
E, L, and F are on the same line. 11) ∠FLO = 90° 12) ∠OLE = 90° 13) ∠FLE = ∠FLO + ∠OLE 14) ∠FLE = 90° + 90° = 180° 15) **Therefore F, L, E are on the same line.**	Angle of the square LOPF. Angle of the square LOME. Addition of angles. Substituting angles by their values. The sides of a straight angle (∠ FLE) are collinear.
L is the midpoint of $\overline{EF}$ 16) FE = FL + LE 17) FL = 3cm 18) LE = 3cm 19) FL = LE 20) F, L, and E are aligned. 21) **Therefore L is the midpoint of $\overline{FE}$.**	Adding segments. Side of the square FLOP. Side of the square LOME. Segments with same measurement (3cm). See Step 15. The point on a segment dividing it into two congruent segments is its midpoint.
Nature of FEMP? 22) $\overline{EM}$ is parallel to $\overline{LO}$ and EM = LO.	Opposite sides of a square (LOME) are parallel and congruent.
23) $\overline{FP}$ is parallel to $\overline{LO}$ and FP = LO.	Opposite sides of a square (LOPF) are parallel and congruent.

Step	Reason
24) $\overline{EM}$ is parallel to $\overline{FP}$	If two lines are parallel to a third one, then they are also parallel (both are parallel to $\overline{LO}$).
25) EM = FP	Transitive property (both are equal to $\overline{LO}$).
26) Then FEMP is a parallelogram	A quadrilateral with two opposite sides that are at the same time parallel and congruent is a parallelogram.
27) But ∠FPL = 90°	Any angle of a square (FLOP) is a right angle.
28) **Therefore FEMP is a rectangle.**	A parallelogram with one right angle is a rectangle.

Perimeter of FEMP

Step	Reason
29) MP = 6cm ; EM = 3cm	Given
30) **The perimeter of FEMP = 2 × (6cm+ 3cm) = 18cm**	Sum of the sides of the rectangle.

5.3 Draw a ¼ circle $\overparen{QL}$ with radii $\overline{OQ}$ and $\overline{OL}$, where OL = 4cm. M is a point on arc $\overparen{QL}$, $\overline{MN}$ is perpendicular to $\overline{OL}$, $\overline{MP}$ is perpendicular to $\overline{OQ}$. Calculate PN.

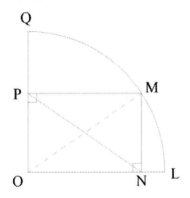

Given:
$\overparen{QML}$ is a ¼ circle with center O and radius OL = 4cm.
$\overline{MN}$ is a perpendicular line to $\overline{OL}$; $\overline{MP}$ is a perpendicular line to $\overline{OQ}$.
Conclusion:
Find the value of PN.
Demonstration:

Step	Reason
1) $\overparen{LMQ}$ = ¼ circle	Given
2) ∠LMQ is congruent to ∠LOQ.	The central angle has the same measurement as the intercepted arc.
3) ∠LOQ = 360° ÷ 4	The central angle of a circle measures 360°.

Step	Reason
4) ∠LOQ = 90°	Calculation
5) $\overline{MN}$ is perpendicular to $\overline{OL}$.	Given
6) ∠MNO = 90°	Two perpendicular lines form a right angle.
7) $\overline{MP}$ is perpendicular to $\overline{OQ}$.	Given
8) ∠MPO = 90°	Two perpendicular lines form a right angle.
9) **MNOP is a rectangle.**	A quadrilateral with three right angles is a rectangle.
Measurement of PN.	
10) PN = OM	The diagonals of a rectangle are congruent.
11) OM = 4cm	Radius of the circle.
12) **Therefore PN = 4cm**	Transitive property.

5.4 Draw a parallelogram ABCD such that ∠ABC = 78°. $\overrightarrow{DI}$ is the angle bisector of ∠ADC, and $\overrightarrow{AJ}$ the angle bisector of ∠DAB. $\overrightarrow{DI}$ and $\overrightarrow{AJ}$ intersect at point M. AB = 7cm and BC = 4.5cm. E is a point on $\overleftrightarrow{AB}$ exterior to ABCD such that AE = 2cm.

a) Find the measurement of angles ∠ADC, ∠DAE, and ∠DAB.

b) Find the measurement of angles ∠MDA and ∠DAM. Deduct the nature of triangle DMA and triangle AMI.

c) Draw the angle bisector of ∠BCD cutting $\overleftrightarrow{DM}$ at N, and the angle bisector of ∠ABC cutting $\overleftrightarrow{CN}$ at P and $\overleftrightarrow{AM}$ at Q.

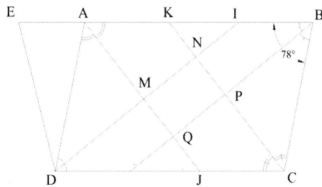

Given:
ABCD is a parallelogram
∠ABC = 78°
$\overrightarrow{DI}$ is the angle bisector of ∠ADC; $\overrightarrow{AJ}$ is the angle bisector of ∠ DAB
Conclusion:
Find the measurements of angles ∠ADC, ∠DAE and ∠DAB.

Demonstration:

Step	Reason
The measurement of ∠ADC 1) ∠ADC = ∠ABC 2) ∠ABC = 78° 3) **Therefore ∠ADC = 78°**	In a parallelogram opposite angles are congruent. Given Transitive property.
The measurement of ∠DAE 4) ∠ADC = 78° 5) $\overrightarrow{AB}$ is parallel to $\overleftrightarrow{DC}$. 6) **Therefore ∠DAE = ∠ADC = 78°**	See Step 3. In a parallelogram, opposite sides are parallel. Alternate interior angles formed by two parallel lines ($\overleftrightarrow{AB}$ and $\overleftrightarrow{DC}$) and a transversal ($\overleftrightarrow{AD}$) are congruent.
The measurement of ∠DAB 7) ∠DAB = 180° - ∠ADC 8) ∠ADC = 78° 9) **Therefore ∠DAB = 180° − 78° = 102°**	In a parallelogram consecutive angles are supplementary. See Step 3. Substituting ∠ADC by its value.
The measurement of ∠MDA 8) $\overrightarrow{DI}$ is the angle bisector of ∠ADC. 10) ∠MDA = ∠ADC ÷ 2 11) ∠ADC= 78° 12) **Therefore ∠MDA = 78° ÷ 2 = 39°**	Given An angle bisector ($\overrightarrow{DI}$) cuts an angle (∠ADC) into two congruent angles. Given
The measurement of ∠DAM 13) ∠DAM = ∠DAB ÷ 2 14) **Therefore ∠DAM = 102° ÷ 2 = 51°**	The angle bisector ($\overrightarrow{AJ}$) cuts the angle (∠ DAB) into two congruent angles. Substituting ∠DAB by its value (See Step 9).
The nature of triangle DMA and triangle AMI. 15) ∠DAM + ∠ MDA = 39° + 51° =	Addition of angles

Step	Reason
90°	
16) **Therefore DMA is a right triangle.**	If two angles of a triangle are complementary then it is a right triangle.
17) ∠DMA = 180° – (∠DAM + ∠MDA) = 180° – 90° = 90°	The sum of the angles of a triangle (DMA) is 180°.
18) ∠DMI = 180°	A line ($\overrightarrow{DI}$) is a straight angle.
19) ∠AMI = ∠DMI - ∠DMA	Subtracting angles.
20) ∠AMI = 180° - 90° = 90°	Substituting angles by their values.
21) **Therefore AMI is a right triangle.**	A triangle with a right angle is a right triangle.
The nature of MNPQ	
22) $\overrightarrow{AJ}$ is the angle bisector of ∠DAB.	Given
23) ∠DAM = ∠MAB	An angle bisector ($\overrightarrow{AJ}$) cuts an angle (∠DAB) into two congruent angles.
24) ∠DAM = 51°	See Step 14.
25) ∠MAB = 51°	Transitive property.
26) ∠AIM = 90° – ∠MAB	In a right triangle (AMI), the acute angles are complementary.
27) ∠AIM = 90° – 51° = 39°	Substituting an angle (∠MAB) by its value.
28) ∠ABP = ∠ABC ÷ 2	An angle bisector ($\overrightarrow{BP}$) cuts an angle (∠ABC) into two congruent angles
29) ∠ABC = 78°	Given
30) ∠ABQ = 78° ÷ 2 = 39°	Substituting an angle (∠ABC) by its value.
31) $\overrightarrow{ID}$ is parallel to $\overrightarrow{BQ}$	If two lines ($\overrightarrow{ID}$ and $\overrightarrow{BQ}$) are cut by a transversal ($\overleftrightarrow{AB}$) forming congruent corresponding angles (∠AIM and ∠ABQ), THEN the lines are parallel.
32) Let K be the intersection point of $\overleftrightarrow{PN}$ and $\overleftrightarrow{AB}$.	Assumption
33) $\overrightarrow{CK}$ is the angle bisector of ∠BCD.	Given
34) ∠BCK = ∠BCD ÷ 2	An angle bisector ($\overrightarrow{CK}$) cuts an angle (∠BCD) into two congruent angles.
35) ∠BCD = 180° – ∠ABC	In a parallelogram, consecutive angles are supplementary.

Step	Reason
36) ∠BCD = 180° − 78° ∠BCD = 102°	Substituting the angle by its value (See Step 28).
37) ∠BCK = ∠BCD ÷ 2 ∠BCK = 102° ÷ 2 = 51°	See Step 33. Substituting the angle by its value.
In triangle BKC	
38) ∠BKC = 180° − (∠CBK +∠BCK)	The sum of the angles of a triangle is 180°.
39) ∠BKC = 180° − (78° + 51°) ∠BKC = 180° − 129° = 51°	Substituting the angles by their values (note: ∠CBK = ∠ABC = 78°).
40) $\overrightarrow{AJ}$ is the angle bisector of ∠ DAB	Given
41) ∠BAM = ∠DAB ÷ 2	An angle bisector ($\overrightarrow{AJ}$) cuts an angle (∠ DAB) into two congruent angles.
42) ∠DAB = ∠BCD = 102°	In a parallelogram, opposite angles are congruent.
43) ∠BAM =102° ÷ 2 = 51°	Substituting the angle by its value.
44) $\overleftrightarrow{MQ}$ is parallel to $\overleftrightarrow{NP}$	If two lines ($\overleftrightarrow{AJ}$ and $\overleftrightarrow{KC}$) are cut by a transversal ($\overleftrightarrow{AB}$) forming congruent corresponding angles (∠BAM and ∠BKC), THEN the lines are parallel.
45) MNPQ is a parallelogram	A quadrilateral with opposite parallel sides is a parallelogram.
46) $\overleftrightarrow{AM}$ is perpendicular to $\overleftrightarrow{DI}$	The sides of a right angle (∠DMA = 90°) are perpendicular. See Step 16.
47) Therefore, MNPQ is a rectangle.	A parallelogram with a right angle is a rectangle.

5.5 Draw a rhombus ABCD with center I such that AB = 5cm and ∠CAB = 60°. Place E the midpoint of $\overline{BC}$. Place J the symmetric point of point I with respect to E.

a. Explain the construction.

b. Demonstrate that IBJC is a rectangle.

c. Find the values of the angles of triangle IBE.

Study Drawing:

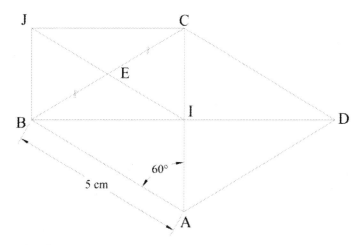

Finding how to draw the figure:

In a rhombus the diagonals are angle bisectors; since ∠CAB = 60°, then ∠CAD = 60°

Furthermore, in a rhombus, diagonals intersect at their midpoint and are perpendicular.

Explanation of the figure:

a. Draw ∠BAx = 60°, then ∠xAD = 60°.

b. Draw $\overline{AB}$ = 5cm

c. Draw $\overline{AD}$ = 5cm

d. Draw $\overline{BD}$ and its midpoint I.

e. Draw $\overline{BC}$ then $\overline{DC}$.

f. Draw $\overline{AI}$ and prolong to C such that IA = IC

Exact Construction:

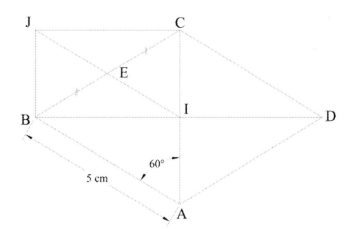

Given:

ABCD is a rhombus with center I

$\angle CAB = 60°$

E is midpoint of segment $\overline{BC}$

J is the symmetric point of I with respect to E

Conclusion:

Prove that IBJC is a rectangle.

Demonstration:

Step	Reason
1) E is the midpoint of $\overline{BC}$	Given
2) J is the symmetric point of I with respect to E.	Given
3) E is the midpoint of $\overline{IJ}$	Two points (I and J) are symmetric with respect to a point (E), if this point (E) is the midpoint of the segment ($\overline{IJ}$).
4) **Therefore IBJC is a parallelogram**	Any quadrilateral with diagonals intersecting in their midpoint is a parallelogram.
IBJC is a rectangle. 5) $\overline{BD}$ and $\overline{AC}$ are perpendicular.	The diagonals in a rhombus are perpendicular.
6) $\angle BIC = 90°$	Two perpendicular lines form a right angle measuring 90°.
7) **Therefore IBJC is a rectangle**	Any parallelogram with a right angle is a rectangle.
The values of the angles of triangle IBE. 8) $\overline{AC}$ is perpendicular to $\overline{BD}$.	The diagonals of a rhombus are perpendicular.
9) ABI is a right triangle.	A triangle with two perpendicular sides is a right triangle.
10) $\angle ABI = 90° - \angle BAI$	In a right triangle, acute angles are complementary
11) $\angle BAI$ (or $\angle BAC$) = 60°	Given
12) $\angle ABI = 90° - 60° = 30°$	Substituting an angle ($\angle BAI$) by its value.
13) $\angle EBI = \angle ABI$	In the rhombus (ABCD) the diagonals ($\overline{BC}$) are angle bisectors ($\angle ABC$).
14) **Therefore $\angle EBI = 30°$**	Transitive property.
15) BE = EC = EJ = EI	In a rectangle (IBJC) the diagonals are congruent and intersect at their midpoint.
16) BEI is an isosceles triangle.	Any triangle with two congruent angles is an isosceles triangle.

Step	Reason
17) ∠EBI = ∠EIB	In an isosceles triangle the base angles are congruent.
18) ∠EBI = 30°	See Step 14.
19) ∠EIB = 30°	Transitive property.
20) ∠BEI = 180° − (∠EBI +∠EIB)	The sum of the angles of a triangle is 180°.
21) ∠BEI = 180° − (30° + 30°)	Substituting the angles by their values.
22) ∠BEI = 180° − 60° = 120°	

5.6 Draw a rectangle ABCD with center I such that AB = 6 cm and ∠BAC = 40°. Draw E and F, as the midpoints of $\overline{BC}$ and $\overline{DC}$ (respectively). J is the symmetric point of I with respect to E, and K is the symmetric point of I with respect to F.

c. Demonstrate that IBJC is a rhombus.

d. Demonstrate that K, C, and J are aligned, and that C is the midpoint of segment $\overline{KJ}$.

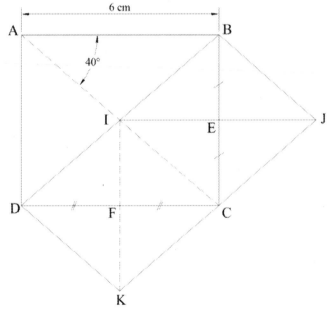

Given:

ABCD is a rectangle with center I

AB = 6cm

∠BAC = 40°

E is the midpoint of $\overline{BC}$ and F is the midpoint of $\overline{DC}$

J = S$_E$ (I) and K = S$_F$ (I)

Conclusion:

Prove that IBJC is a rhombus.

Demonstration:

Step	Reason
1) AC = BD	In a rectangle (ABCD) the diagonals are congruent.
2) IB = IC	In a rectangle the diagonals intersect at their midpoint.
3) Then IBC is an isosceles triangle.	Any triangle with two congruent sides (IB and IC) is an isosceles triangle.
4) E is the midpoint of $\overline{BC}$	Given
5) $\overline{IE}$ is the altitude, from I, of triangle IBC.	In an isosceles triangle, the median from the vertex angle is at the same time the altitude.
6) Then $\overline{BC}$ and $\overline{BE}$ are perpendicular to $\overline{IE}$ and $\overline{IJ}$.	The altitude of a triangle is the perpendicular line of a side passing through the opposite vertex.
7) E is the midpoint of $\overline{IJ}$.	If two points (I and J) are symmetric with respect to a point (E), then this point (E) is the midpoint of the segment ($\overline{IJ}$).
8) $\overline{BE}$ is one of the three **medians** of triangle IBJ.	The median of a triangle is the ray joining a vertex to the midpoint of the opposite side. (Chapter 1)
9) $\overline{BE}$ is perpendicular to $\overline{IJ}$.	See Step 6.
10) $\overline{BE}$ is one of the **altitudes** of triangle IBJ.	The altitude is the perpendicular line of a side, which passes through the opposite vertex.
11) Then IBJ is an isosceles triangle.	If the median of a triangle is at the same time its altitude, then it is an isosceles triangle.
12) BI = BJ	The legs of an isosceles triangle are congruent.
13) E is the midpoint of $\overline{BC}$.	Given
14) BE = EC	A midpoint divides a segment into two congruent segments.
15) E is the midpoint of $\overline{IJ}$.	See Step 7.
16) Then IBJC is a parallelogram.	Any quadrilateral with diagonals intersecting at their midpoint is a parallelogram.
17) **Therefore, IBJC is a rhombus.**	Any parallelogram with two congruent and consecutive sides (BI and BJ) is a rhombus.
K, C, and J are collinear. 18) AI = IB	The diagonals in a rectangle (ABCD) are congruent and intersect at their midpoint.

Step	Reason
19) AIB is an isosceles triangle.	Any triangle with two congruent sides is an isosceles triangle.
20) $\angle$IAB= 40°	Given
21) Then $\angle$IBA = $\angle$IAB = 40°	The base angles of an isosceles triangle (AIB) are congruent.
22) IBC is an isosceles triangle.	See Step 3.
23) Then $\angle$IBC = $\angle$ICB	The base angles of an isosceles triangle (IBC) are congruent.
24) $\angle$IBC = $\angle$ABC – $\angle$IBA	Subtraction of angles.
25) $\angle$ABC = 90° and $\angle$IBA = 40°	Given and see Step 18.
26) Then $\angle$IBC = 90° – 40° = 50°	Substituting angles by their values.
27) $\angle$ IBJ = 2 × $\angle$IBC	In a rhombus the diagonals are angle bisectors.
28) $\angle$ IBJ = 2 × 50° = 100°	Substituting the angle $\angle$IBC by its value.
29) $\angle$ICJ = $\angle$ IBJ = 100°	In a rhombus opposite angles are congruent.
30) $\angle$ICD = $\angle$BCD – $\angle$ BCI	Subtraction of angles.
31) $\angle$BCD= 90°	Any angle of a rectangle is a right angle.
32) And $\angle$ BCI = $\angle$IBC = 50°	See Steps 23 and 26.
33) $\angle$ICD = 90° – 50° = 40°	Substituting angles by their values.
34) ICKD is also a rhombus.	Same Reasons as steps 1-15 for IBJC.
35) $\angle$ICK = 2 ×$\angle$ICD =	The diagonal ($\overline{DC}$) of a rhombus, is one of its angle bisectors (of $\angle$ICK).
36) $\angle$ICK = 2 × 40° = 80°	Substituting the angle by its value.
Therefore	
37) $\angle$JCK = $\angle$ICJ +$\angle$ ICK	Addition of adjacent angles.
38) $\angle$JCK = 100° + 80° = 180°	Substituting the angles by their values from steps 29 and 36.
39) $\angle$JCK is a straight angle, and J, C, and K are collinear.	The sides of a straight angle are on the same line.

5.7 Draw a rectangle ABCD. The perpendicular lines at the endpoints of the diagonals $\overline{AC}$ and $\overline{BD}$ intersect at E, F, G, and H. Demonstrate that EFGH is a rhombus.

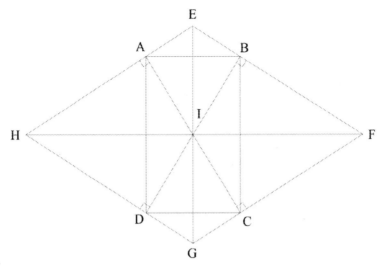

Given:

ABCD is a rectangle.

$\overline{EF}$ and $\overline{GH}$ are perpendicular lines to segment $\overline{DB}$.

$\overline{EH}$ and $\overline{FG}$ are perpendicular to segment $\overline{AC}$.

Conclusion:

Prove that EFGH is a rhombus.

Demonstration:

Step	Reason
EFGH is a parallelogram 1) $\overline{EF}$ and $\overline{GH}$ are perpendicular to $\overline{BD}$.	Given
2) Then $\overline{EF}$ and $\overline{GH}$ are parallel.	If two lines are perpendicular to a third one, then they are parallel.
3) $\overline{EH}$ and $\overline{FG}$ are perpendicular to $\overline{AC}$.	Given
4) $\overline{EH}$ and $\overline{FG}$ are also parallel.	If two lines are perpendicular to a third one, then they are parallel.
5) Then EFGH is a parallelogram.	Any quadrilateral with parallel opposite sides is a parallelogram.
EFGH is a rhombus 6) AI = AC ÷ 2 and BI = BD ÷ 2	The diagonals in a rectangle intersect at their midpoints.
7) AC = BD AC ÷ 2 = BD ÷ 2	The diagonals of a rectangle are congruent.

Step	Reason
8) AI = BI	Substitution
9) $\overline{AI}$ is perpendicular segment to $\overline{EH}$.	Given
10) $\overline{BI}$ is perpendicular to $\overline{EF}$.	Given
11) Then AEI and BEI are right triangles.	Any triangle with a right angle is a right triangle.
12) $\overline{EI}$ is the hypotenuse for triangles AEI and BEI.	The hypotenuse is the opposite side of the right angle.
13) Then triangles AEI and BEI are congruent.	Two right triangles with the same hypotenuse and one congruent leg are congruent.
14) And AE = BE	Corresponding sides of congruent triangles are congruent.
15) HI = IF	In a parallelogram (EFGH) the diagonals intersect at their midpoint (I).
16) And triangles HAI and FBI are congruent.	Same Reasons as Steps 6-13.
17) Then HA = FB	Corresponding sides of congruent triangles are congruent.
18) HA + AE = HE and FB + BE = FE	Segments formed by addition of congruent segments.
19) HE = FE	Substitution
20) **Therefore EFGH is a rhombus.**	Any parallelogram with two consecutive congruent sides is a rhombus.